ADVANCED MULTIPI
CHEMISTRY (with answe

J.A.S. REES, B.Sc., L.R.I.C.

Senior Science Master,
Belmont Abbey

HARRAP LONDON

By the same author
Multiple-choice Chemistry

In this series
Advanced Multiple-choice Biology
R. Soper and S.T. Smith

First published in Great Britain 1975
by GEORGE G. HARRAP & CO. LTD
182-184 High Holborn, London WC1V 7AX

ISBN 0 245 52336 7

Printed in Great Britain by
Biddles Ltd Guildford Surrey

PREFACE

This text has been designed for candidates following a modern A Level Course. The questions have been pretested by members of the Sixth Form at Belmont Abbey, Hereford, and at The College, Llandovery. In the light of the results from pretesting it was apparent that the questions could be used profitably, either simply as a source of test questions or as a supplement to theory, by setting questions directly following a theory session for the pupil to complete under teacher supervision, or for the pupil to complete in his or her own time with the aid of reference texts. The book could also be used at the end of the academic course in two ways, either again as a source of tests, each test being taken as a whole, or by selecting a single question from a series of tests in order to determine the pupil's overall ability – e.g., Questions 4 and 5 from Sections 1-12 inclusive.

I am sincerely grateful to Dr G. H. Goddard at The College, Llandovery, M. A. Lloyd, B.Sc. at The Grammar School, Ross-on-Wye and A. Firman, M.A. at Cross Green School, Leeds for their assistance in pretesting and reviewing the items contained in this text.

1975 J. A. S. R.

Reference Texts

Heys, H. L. *New Organic Chemistry* (Harrap)
Heys, H. L. *Physical Chemistry* (Harrap)
Holderness *Inorganic and Physical Chemistry* (Heinemann)
Moody, B. J. *Comparative Inorganic Chemistry* (E. Arnold)
Muir, T. A. and Lambert, J. *Practical Chemistry* (Heinemann)
Murray, P. R. S. *Principles of Organic Chemistry* (Heinemann)
Nuffield Advanced Science: Student Books 1 and 2, Teacher's Guides 1 and 2, and Book of Data
Vogel, A. I. *Textbook of Macro and Semimicro Qualitative Inorganic Analysis* (Longman)
Waddington, D. J. *Organic Chemistry* (Mills and Boon)
Waddington, D. J. and Finlay, H. S. *Organic Chemistry Through Experiment* (Mills and Boon)

KEY

(*a*) **Five-choice completion**

These are items which are followed by five responses and you must select the response which you think is correct.

(*b*) **Multiple selection**

These are items which are followed by four statements and you must select those statements which you think are correct from the possibilities listed.

Response	*Correct statements*
A	1, 2 and 3
B	1 and 3
C	2 and 4
D	4 only
E	ANY OTHER combination

(*c*) **Items consisting of two statements**

The item consists of two statements which may or may not be directly related and you must select a response from the list given below which you think applies to the statements.

Response	STATEMENT 1	STATEMENT 2
A	True	True, and explains Statement 1 correctly
B	True	True, but is independent of Statement 1
C	True	False
D	False	True
E	False	False

Pupil Requirements

Access to a data book, preferably the one produced by the Nuffield Foundation, and a copy of a Periodic Table; these items *may* be required by the pupil depending upon the depth to which the pupil had studied the course.

CONTENTS

1 RADIOACTIVITY, STRUCTURE OF THE ATOM, THE MOLE

Questions 1 to 5 refer to the responses set out below. Select the response which you believe to answer the question correctly.

A isotope **C** neutron **E** proton
B α-particle **D** mole

Which of the above

1 has a mass number of 4?

2 contains the same number of elementary units as there are atoms in 12 g of carbon-12?

3 has a mass of 1 atomic mass unit and charge equivalent in size to that of an electron?

4 is used to bring about the fission of Uranium-235?

5 is the name for an element which has similar chemical properties to, but different physical properties from, another element which has the same atomic number?

6 Which of the following is determined by means of a mass spectrometer?

A Avogadro's constant
B number of protons in an atom
C the existence of isotopes
D the frequency of X-rays emitted from an element
E the number of electrons present in an energy level

7 When a parallel beam of α-particles is incident at a thin gold foil most of the particles pass through, but some are deviated because

A α-particles are not charged
B electrons are present around the gold nucleus
C gold has a high density
D the nucleus of the gold atom has a positive charge
E different α-particles have different momenta

8 Which of the following could be determined by means of the apparatus shown in the diagram?

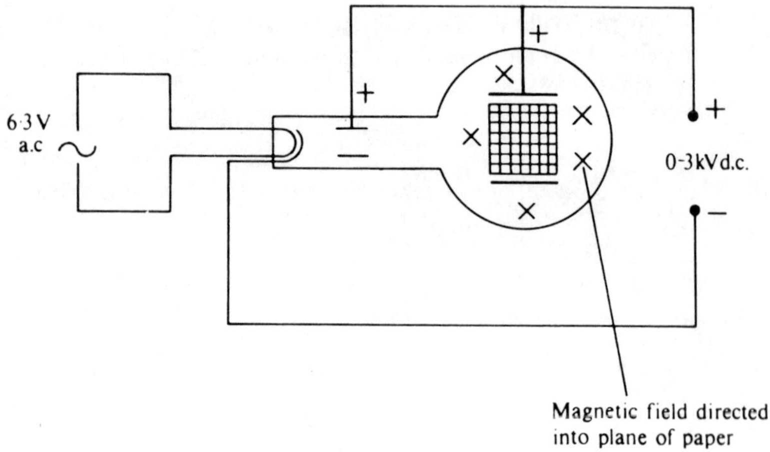

A the magnitude of charge on an electron
B the ratio of charge to mass for an electron
C the ionization energy of an inert gas
D the quantity of charge transmitted by a current of 1 ampere
E the relation between the temperature of the filament and the amount of thermionic emission

9 Which of the following is true of the element lithium, ${}^{7}_{3}Li$?
A lithium is a metal
B the element is radioactive
C the atom has a mass number 10
D the element will form a cation with charge +3
E 4 protons are present in the nucleus

10 A drop of a dilute solution of a non-volatile organic compound in a volatile solvent was placed on a clean water surface. A film was observed on the water surface. Which of the following quantities are necessary to determine an approximate value of the size of one molecule of the non-volatile compound?

	mass of 1 mole of organic compound	*area of film*	*volume of drop of solution*
A	x		x
B	x	x	
C			x
D		x	x
E	x		x

11 Why was a clean water surface required in the above experiment?

1 to lower the surface tension of the solution
2 to enable an approximately circular film to form
3 to ensure that the film was clearly visible
4 to ensure that the non-volatile compound did not undergo a chemical change

12 Which of the following is it possible to determine if a known quantity of charge is passed through an electrolyte and a known number of moles of a Group 1 element is deposited at the cathode of an electrolytic cell?

1 the charge on an electron
2 Avogadro's Constant
3 the conductivity of the electrolyte
4 the Faraday Constant

13 Which of the following apply to the nucleus of a helium atom?

1 2 neutrons are present
2 it is approximately 8 000 times heavier than an electron
3 it requires 2 Faradays of charge per mole for complete reduction
4 it is a product of thermionic emission

14 Which of the following occur when an atom $^{238}_{92}U$ undergoes radioactive decay to form $^{234}_{92}U$?

1 a mass change of 4 amu
2 the emission of 2 β-particles
3 an emission of energy
4 nuclear fusion

15 Which of the following apply to the reaction given below?

$$^{2}_{1}H + ^{3}_{1}H \rightarrow ^{4}_{2}He + ^{1}_{0}n; -\Delta H$$

1 ΔH is a quantity of energy
2 the reaction is an example of nuclear fission
3 $^{1}_{0}n$ is an uncharged particle
4 2 atoms of $^{2}_{1}H$ are identical with one atom of $^{4}_{2}He$

	STATEMENT 1	STATEMENT 2
16	Standard temperature is 273 K and standard pressure is 1 atmosphere.	At standard temperature and pressure 1 mole of any gas occupies 22·4 dm^3
17	The ratio of the mass of 1 mole of gas to the mass of 1 mole of hydrogen is the relative molecular mass of the gas.	The mass of 1 mole of hydrogen is 2 g.

	STATEMENT 1	STATEMENT 2
18	γ-radiation has a high velocity of propagation.	$E = mc^2$.
19	1 mole of phosphorus molecules (P_4) contains Avogadro's number of phosphorus atoms.	1 mole of calcium chloride contains 1 mole of chloride ions.
20	The mass of 1 carbon atom, ${}^{12}_{6}C$, is equivalent to six units of atomic mass.	The ${}^{12}_{6}C$ atom contains 6 protons.
21	The ratio of the number of atoms of nitrogen to oxygen in nitrogen monoxide [nitrogen(I) oxide] is 1:1.	The type of molecule formed when two gases combine is dependant upon the number of molecules of each gas present only and independent of the physical conditions.

22 Which of the following apply to the statement that the helium isotope, ${}^{6}_{2}He$, has a half-life of 0·80 seconds?

A in 0·80s 1 atom of He is formed
B the ${}^{6}_{2}He$ atom emits high velocity neutrons
C the ${}^{6}_{2}He$ isotope is a source of γ-radiation
D one half of the original number of He atoms will remain after 0·80s
E the relation of $E = mc^2$ applies to the disintegration of the ${}^{6}_{2}He$ nucleus

23 Which of the following is it necessary to know in order to determine the mass of 1 mole of a gas produced during a chemical reaction?

1 volume of gas
2 temperature at which gas was collected
3 pressure of gas
4 mass of gas collected

24 The apparatus shown can be used to demonstrate all of the following EXCEPT that cathode rays

A are produced by thermionic emission
B can cause fluorescence of some materials
C travel in straight lines
D will undergo diffraction
E are absorbed by a metal

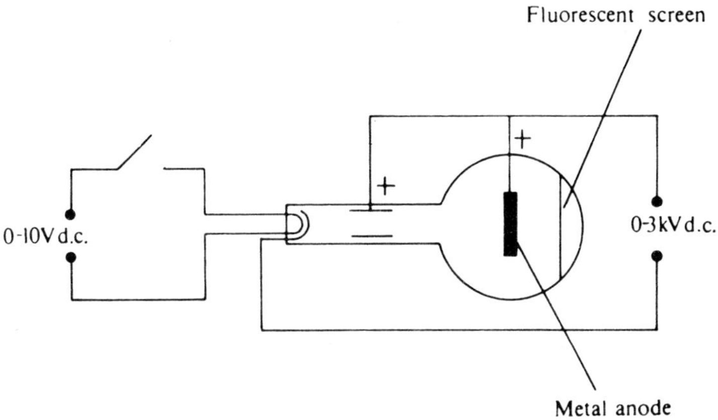

25 Which of the following is an indication of the stability of the nucleus of an atom?

A the temperature at which thermionic emission occurs

B the ease with which ionization occurs

C the position of the element in the Electrochemical Series (Activity Series)

D the ratio of the number of protons to the number of neutrons present in the atom

E the difference between the total mass of the individual protons and neutrons present in the nucleus and the experimentally determined atomic mass

2 IONIZATION ENERGIES, SPECTRA, PERIODICITY

1 Which of the following elements has the *least* value of 1st ionization energy?

A hydrogen **C** potassium **E** sulphur
B chlorine **D** chromium

2 Which of the following spectral series occurs in the visible region of the electromagnetic spectrum?

A Pfund **C** Paschen **E** Lyman
B Brackett **D** Balmer

3 Each of the following is evidence of the increasing non-metallic character of the elements with increasing atomic number in any Period EXCEPT

A decreasing atomic radius
B increase in atomic number
C decrease in boiling point
D increase in volatility of chlorides
E increase in acidic character of oxides

4 Which of the following is characteristic of the gradation in properties of the elements when descending a Group of the Periodic Table?

A increase in electronegativity
B increase in tendency to ionize by electron loss
C decrease in ionic radius
D decrease in tendency to form complex ions
E decrease in crystalline nature of the elements

5 The chlorides of the common elements in Period 3 are said to exhibit periodicity because

A ionic chlorides are formed on combination of the element and chlorine
B the chlorides of non-metals are insoluble in water
C the pH of aqueous solutions of the chlorides increases across the Period
D aluminium chloride has a lattice layer structure
E 2 moles of chlorine combine with 1 mole of iron in an endothermic reaction

6 The first three ionization energies, in $kJ\,mol^{-1}$, for five elements are listed below, which of the elements can be isolated by electrolysis of the fused hydroxide?

A	420	3 100	4 400
B	1 260	2 300	3 800
C	1 520	2 700	3 900
D	1 090	2 400	4 600
E	710	1 400	2 900

7 Which of the following elements has the *highest* melting point?

A lithium
B caesium
C calcium
D zinc
E lead

8 Which of the following elements will have the *greatest* latent heat of fusion?

A sodium
B magnesium
C aluminium
D silicon
E phosphorus (red)

9 Each of the following pairs of elements have similar latent heats of vaporization EXCEPT

A	beryllium	aluminium
B	lithium	magnesium
C	iron	cobalt
D	silver	mercury
E	argon	fluorine

10 Which of the following elements will be the *least* electronegative?

A aluminium
B copper
C mercury
D silver
E tin

11 Which of the following is described by a quantum number?

1 an energy level containing electrons
2 the wavelength of a spectral line
3 the promotion of an electron from one orbit to another
4 the energy emitted from an atom according to the relation $E = h\nu$

12 Which of the following will apply to the elements of any Group of the Periodic Table?

1 increase in electropositive character with increase in atomic number
2 all have the same number of energy levels
3 all form ions with the same oxidation number
4 all give the same emission spectra

13 Which of the following apply to an element which has values for the first three ionization energies as shown?

1st	2nd	3rd	
730	2 100	3 400	(kJ mol^{-1})

1 the element is in Group 1
2 an ion of oxidation number 1 can be formed
3 the radius of the cation is small
4 the radius of the atom is larger than the radius of the cation

14 Which of the following may be deduced from the graph of wavelength, λ, of line spectra against n^2, where n = quantum number plus an integer? The element for which the graph was plotted is in Group 1.

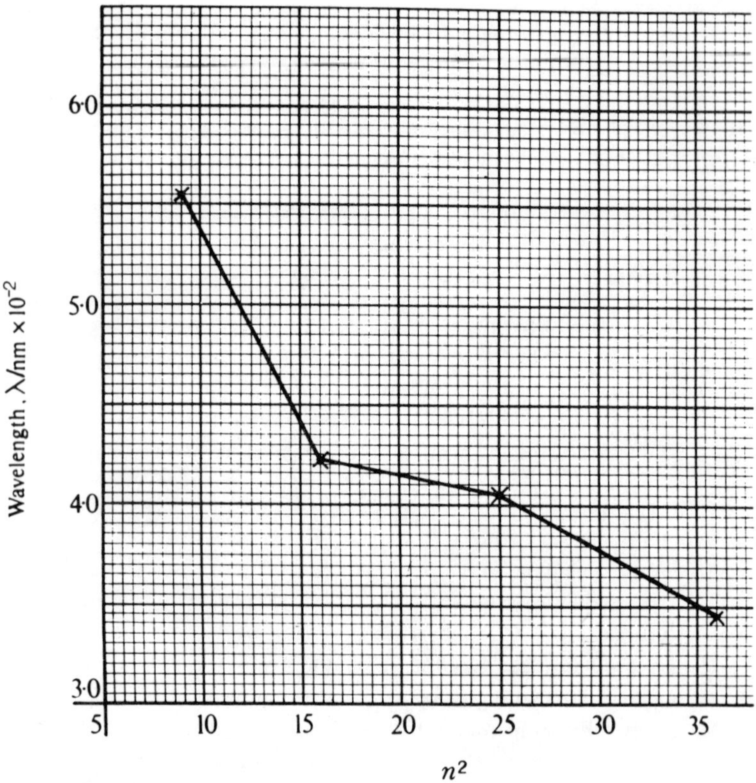

1 one spectral line is in the uv region of the spectrum
2 the line spectra correspond to 4 energy levels
3 the halide will colour a bunsen flame red
4 λ is not proportional to n^2 for a given series

15 Which of the following can be deduced from the energy level diagram given below?

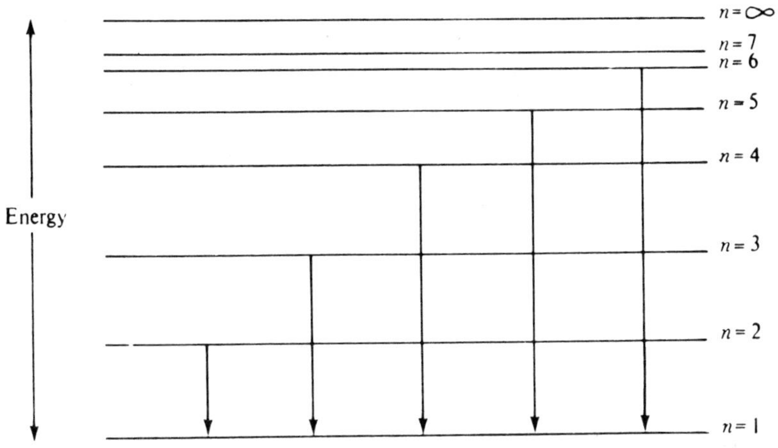

1 5 electrons surround the nucleus of the atom
2 spectral lines are observed in the uv region of the electromagnetic spectrum
3 the diagram refers to an element in Group 2
4 energy is emitted if an electron is transferred from the $n = 5$ to the $n = 1$ level

	STATEMENT 1	STATEMENT 2
16	The elements of Group 4 are all non-metallic in character.	Carbon is a typical non-metal
17	Phosphorus(III) chloride has a molecular structure	Phosphorus(III) chloride is the halide of a Group 5 element.
18	A sodium atom requires less energy to ionize than an atom of caesium.	The 1st ionization energy of sodium is greater than that of caesium.
19	If a sodium lamp is viewed through a diffraction grating a line emission spectrum is observed.	Electrons are excited and on moving back from a high energy level to a lower level emit light at a particular frequency.
20	Only one electron can be situated in the $n = 1$ level at any given time.	The $n = 1$ level is furthest from the nucleus.

21 Use the graph of electrons removed against logarithm of ionization energy to deduce which of the following does **not** apply to element Q.

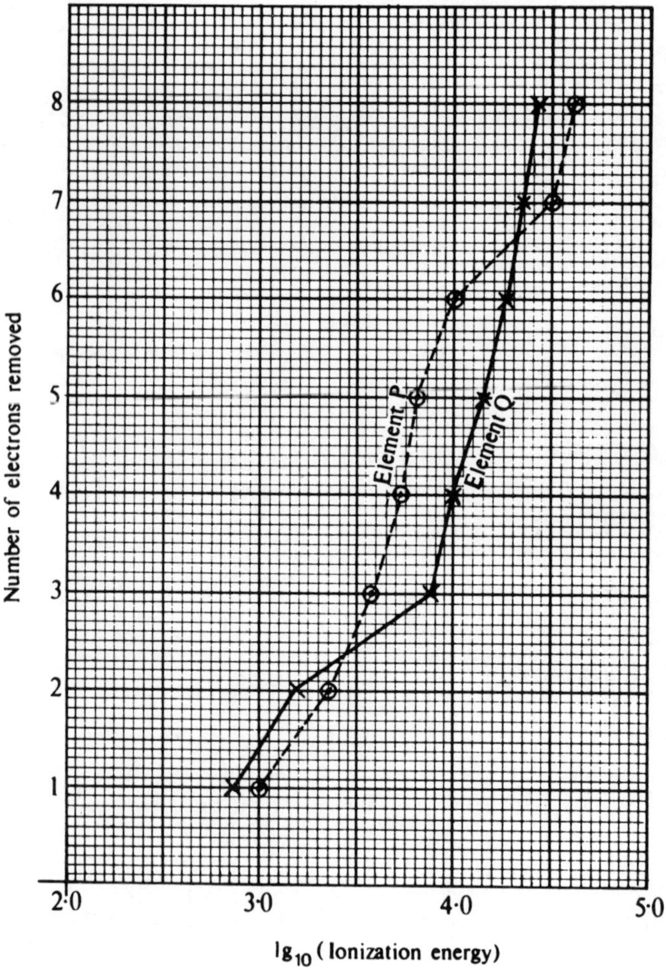

A Q is more metallic than P

B Q can form an ion of oxidation number 2

C less energy is required to remove the 8th electron from P than from Q

D the graph shows the log (ionization energy) for all $n = 1$ and $n = 2$ electrons

E the 3rd electron to be removed is in a different energy level from the 2nd

22 The first three ionization energies (kJ mol^{-1}) of two metals, L and M, are given below. Which of the following can be deduced from these values?

L	650	1 600	3 000
M	760	1 600	3 000

A L has a higher atomic number than M
B the elements will have different chemical properties
C the elements will form salts of type $M^{2+}.SO_4^{2-}$
D ions of oxidation number +1 will occur in aqueous solution
E the principal salts of the elements will be colourless

	STATEMENT 1	STATEMENT 2
23	1 joule of work is done when a potential difference of 1 volt is applied to a gas and 1 coulomb of charge flows through the gas.	An electron of charge $1{\cdot}6 \times 10^{-19}$ C has $1{\cdot}6 \times 10^{-19}$ J of work done on it when it moves between two points which are at a potential difference of 1 volt.

24 Which of the following change when moving from an element in one Period to another element in the same Group but in the next Period?

1 mass number
2 proton number
3 electronic configuration
4 oxidation number of the principal ion

25 The lower elements of any Group may be distinguished from the elements at the top of the Group because the lower elements

A have larger (covalent) radii
B form anions
C are not obtained by reduction of oxides
D form stable hydroxides
E are reducing agents

3 ELECTRONIC CONFIGURATION, IONIZATION ENERGIES, PERIODICITY

1 Which of the following apply to an element for which the graph of energy levels against ionization energies is as shown?

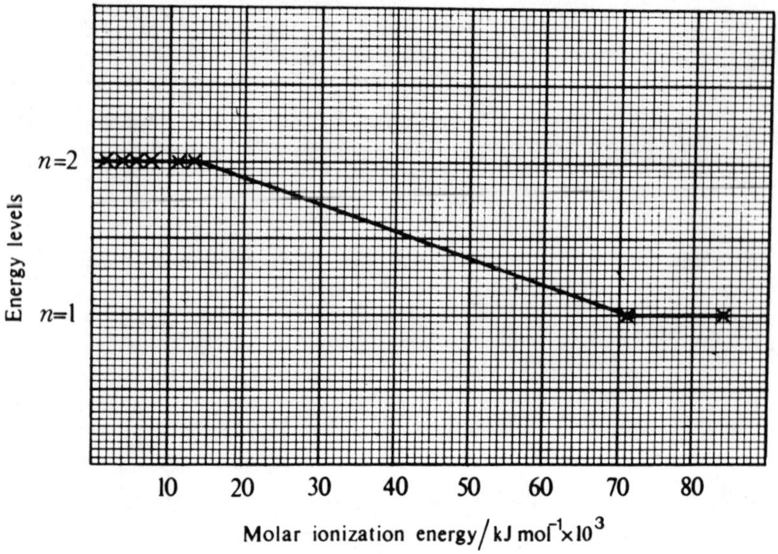

A 6 electrons are easily removed
B the element is an inert gas
C the element is in Group 2
D 2 electrons are present in the 1s orbital
E all the electrons in the $n = 2$ level are unpaired

2 If an atom of an element has the electronic configuration

$1s^2 \quad 2s^2 \quad 2p^6 \quad 3s^2 \quad 3p^1$

then

A the element is a transition element
B the atomic number is 27
C the 1s electrons have the lowest ionization energies
D the electrons in the 2p sub-orbitals are paired
E there are 5 possible energy levels for the electrons to occupy

3 Which of the following is the electronic configuration of chromium (atomic number 24)?

A $1s^2$ $2s^2$ $2p^6$ $3s^2$ $3p^6$ $3d^5$ $4s^1$

B $1s^2$ $2s^2$ $2p^6$ $3s^2$ $3p^6$ $3d^6$

C $1s^2$ $2s^2$ $2p^6$ $3s^2$ $3p^6$ $3d^2$ $4s^4$

D $1s^2$ $2s^2$ $2p^6$ $3s^2$ $3p^2$ $3d^6$ $4s^2$

E $1s^2$ $2s^2$ $2p^6$ $3s^2$ $3p^6$ $3d^3$ $4s^2$ $4d^1$

4 When one atom combines chemically with another atom then each atom attempts to attain an inert gas configuration because

A inert gases do not ionize

B all the inert gases have the same atomic radii

C the inert gases have 2 electrons in the $n = 1$ energy level

D strongly electronegative elements are separated from strongly electropositive elements in the Periodic Table by the inert gases

E energies involved in a chemical reaction only affect electrons furthest from the nucleus and the inert gas configuration is the most stable configuration

5 The electronic configuration of the electrons in the highest energy level for an atom in Group 4 is shown by

A $2s^2$

B $3s^2$ $3p^2$

C $3s^2$ $3p^4$

D $4s^2$ $4p^4$

E $5s^2$ $5p^4$

6 If an atom of an element, X, atomic number 20, combines with 2 atoms of a halogen, Y, atomic number 17, to form an ionic compound, which of the following configurations will be attained by the ions?

	Ion of X	*Ion of Y*
A	argon	argon
B	krypton	argon
C	neon	argon
D	neon	krypton
E	krypton	xenon

7 Ammonia is said to have a 'lone pair' of electrons because

A ammonia is a hydride of a Group 5 element

B ammonia gives rise to a series of ionic salts

C nitrogen has the electronic configuration $1s^2$ $2s^2$ $3p^3$

D nitrogen attains an inert configuration by sharing 3 electrons with 3 hydrogen atoms

E nitrogen still has mutually paired electrons in an outer shell

8 The diagrams set out below show the electrons in the outer orbitals of the reacting elements. Select the reaction in which the configuration is incorrect.

A  1 shared pair

B 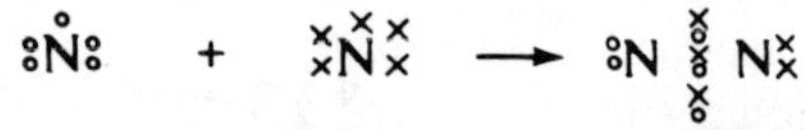 3 shared pairs

C 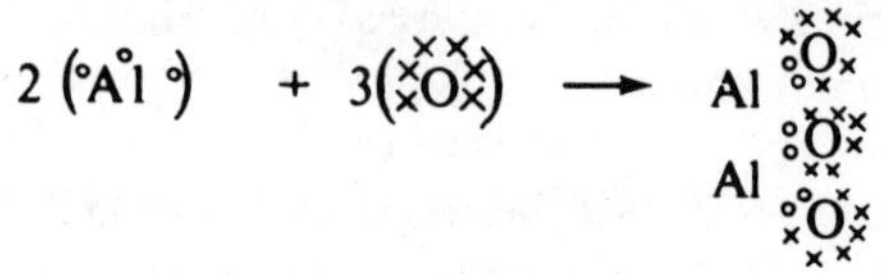 3 pairs donated by the aluminium

D  1 pair donated by oxygen, 2 shared pairs

E  4 shared pairs

9 If all the energy levels from $n = 1$ to $n = 3$ for a particular atom are filled with electrons and 2 electrons are present in the $n = 4$ level which of the following is correct?

A the element is in Group 4
B 30 protons are present in the nucleus
C the electronic configuration is $1s^2\ 2s^2\ 2p^6\ 3s^2$
D the emission spectrum will consist of 4 lines
E 8 electrons are present in each of the $n = 2$ and $n = 3$ levels

10 The first three ionization energies for elements X and Y are as shown.

X	500	4 600	6 900	($kJ\,mol^{-1}$)
Y	590	1 100	4 900	($kJ\,mol^{-1}$)

Why is X different from Y?

A the halides of X are ionic, those of Y are molecular
B X forms an ion X^+, Y forms an ion Y^{++}
C X will displace hydrogen from water, Y will not
D X is a Group 2 element, Y is in Group 7
E X forms an oxide X_2O, Y forms an oxide Y_2O_3

11 Which of the following are typical of a Group 1 element?

1 has complete K and L electron shells
2 forms basic oxides
3 has a low 1st ionization energy
4 forms an ionic hydride of type $M^+.H^-$

12 Which of the following apply to Period 2?

1 an element of atomic number 9 forms an ion with the argon configuration

2 an element in the middle of the Period can exhibit an oxidation number of +6

3 the heat of formation of the halides decreases between elements of atomic number 4 to 9

4 the K electron shell is complete

13 If two ions react to form a compound then the bond between the original ions will have covalent character if

1 the electron clouds from each ion do not overlap in the compound

2 the cation is small compared with the anion

3 the cation has an inert gas configuration

4 the anion is large compared with the ion F^-

14 Which of the following can be deduced from the graph of density against atomic number of the element?

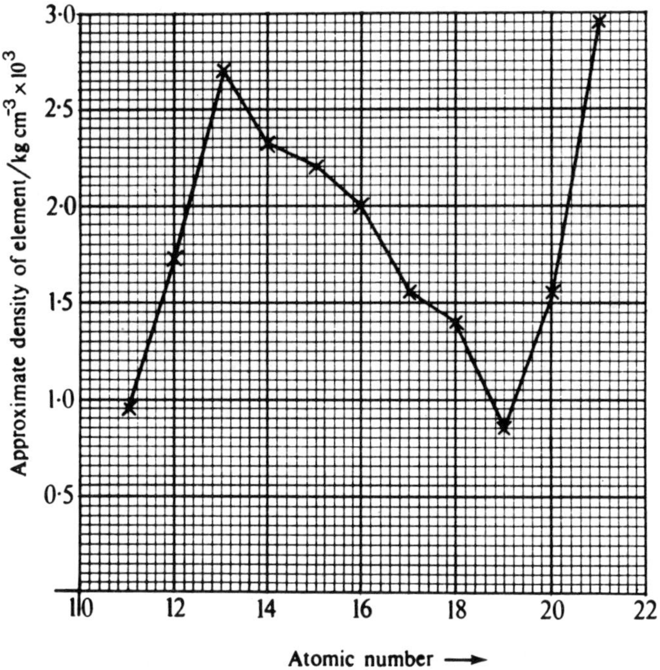

1 the density of the elements exhibit periodicity

2 the covalent radii of the elements exhibit periodicity

3 some Group 1 elements will float on water

4 the element of atomic number 17 is gaseous

15 Which of the following are characteristic of Group 2?

1 the covalent radius of an atom in Period 6 is larger than that of an atom in Period 2
2 the ionic radius decreases from top to bottom
3 oxides of Period 4 will have molecular structures
4 1s, 2s and 2p orbitals of each element is complete

	STATEMENT 1	STATEMENT 2
16	$Mg \rightarrow Mg^{+} + e; +\Delta H$	An atom of a metal can acquire inert gas configuration by electron loss.
17	At room temperature fluorine is less reactive than iodine.	The chemical properties of the elements are a periodic function of the atomic number of the elements.
18	The melting point of silicon will be lower than that of aluminium or phosphorus.	The melting points of the elements exhibit periodicity and silicon is in Group 4.
19	The latent heat of vaporization will be smaller for beryllium than for magnesium.	An atom of beryllium has a smaller volume than an atom of magnesium.
20	The higher the melting point of the element the more likely it is for the atoms to be in a giant lattice structure.	Carbon has a high melting point and the atoms of one allotrope are in a giant lattice structure.

Questions 21 to 25 refer to the list of elements set out below. Select the response which you think to be the correct answer to the question.

A caesium **C** iron **E** bromine
B magnesium **D** aluminium

Which element will

21 attain the krypton configuration upon ionization?

22 have an incomplete 3d sub-orbital?

23 require the least energy in order to form an ion?

24 have one electron in the 3p sub-orbital?

25 have the largest atomic number?

4 s-BLOCK ELEMENTS, Li, Na, K. Ca, Mg, Ba

Questions 1 to 5 refer to the list of elements given below. Select the response which you think is the correct answer to the question.

A lithium **C** magnesium **E** barium
B sodium **D** calcium

Which of the above will

1 require the most energy per mole of element for the removal of the two outermost electrons?

2 have the highest melting point?

3 have the lowest specific density at room temperature?

4 have the electronic configuration?

4s
Ar(core) [↑↓]

5 form the least ionic iodide?

6 Which of the following will **not** yield a metal oxide on strong heating?

A lithium hydroxide
B sodium carbonate
C calcium nitrate
D sodium peroxide
E magnesium sulphate

7 Which of the following will give an aqueous solution of pH = 9?

A lithium hydride
B sodium chloride
C magnesium nitrate
D calcium sulphate
E barium peroxide

8 Which of the following can be used to describe the direct combination between lithium and hydrogen to form lithium hydride?

A $Li + H \longrightarrow Li\text{-}H$

B $2Li + H_2 \longrightarrow 2Li\,H$

C $2Li + H_2 \longrightarrow 2(Li.^+H^-)$

D $Li + e^- \longrightarrow Li^-$
$H - e^- \longrightarrow H^+$

E $Li + H^+ \longrightarrow (Li.H)^+$
$Li + H^- \longrightarrow (Li.H)^-$
$(Li.H)^+ + (Li.H)^- \longrightarrow 2LiH$

9 Which of the following pairs of aqueous solutions will give a white precipitate when mixed at room temperature?

A barium chloride, lead nitrate
B sodium hydroxide, oxalic acid
C sodium thiosulphate, iron (III) chloride
D barium sulphide, hydrochloric acid
E potassium thiocyanate, iron (II) sulphate

10 The compound with the lowest solubility (mol dm^{-3}) at room temperature is

A barium nitrate
B calcium chloride
C magnesium sulphate
D sodium acetate
E lithium carbonate

11 A yellow precipitate was observed when a saturated aqueous solution containing an s-block ion was added to aqueous potassium chromate, and a white precipitate was observed when the solution was added to aqueous ammonium oxalate, the s-block ion was

A Li^+
B Na^+
C Mg^{2+}
D Ca^{2+}
E Ba^{2+}

12 Aluminium potassium sulphate

1 is a double salt
2 may be represented as $K^+ Al^{3+}\ 2SO_4^{2-}.\ 12H_2O$
3 forms octahedral crystals
4 is prepared by the addition of potassium hydroxide to aqueous aluminium sulphate

13 A white precipitate will result when aqueous magnesium sulphate is added to an aqueous solution of

1 sodium hydroxide
2 ammonia
3 sodium carbonate
4 sodium phosphate

14 s-block elements are *all* reducing agents because they

1 only form positively charged ions
2 give an exothermic reaction with water
3 have fixed oxidation numbers
4 have a low 1st ionization energy

15 Which of the following are applicable to calcium?

1 lower melting point than magnesium
2 does not occur free in nature
3 forms a covalent peroxide
4 may be obtained from the chloride by an electrolytic method

	STATEMENT 1	STATEMENT 2
16	The colour observed when sodium chloride is placed in a bunsen flame is yellow.	Electrons emit energy when moving from the 3s to the 3p sub-orbital
17	$2Al(s) + 4OH^-(aq) \rightarrow 2AlO_2^{2-}(aq) + 2H_2(g); -\Delta H$	A 2M aqueous solution of sodium hydroxide has a pH$>$9
18	Potassium carbonate is more stable to heat than sodium carbonate.	The potassium ion is smaller than the sodium ion.
19	Both sodium chloride and magnesium oxide form simple cubical crystals.	Each metal ion is surrounded by 6 non-metal ions in a unit cell of the crystal.
20	Simple salts of s-block elements containing small anions have high solubility.	Sodium fluoride is more soluble than sodium iodide.

21 Why is an acidified solution of potassium permanganate decolourized on the addition of warm concentrated aqueous sodium nitrate?

A oxygen is liberated

B sodium nitrite is an oxidizing agent

C $2MnO_4^-(aq) + 5NO_2^-(aq) + 6H^+(aq) \rightarrow 2Mn^{2+}(aq) + 3H_2O(1) + 5NO_3^-(aq)$

D 1 mole of potassium permanganate is equivalent to 1 Faraday of charge

E nitrogen dioxide is formed which reduces the permanganate

22 Which of the following salts contains an element or an ion with an oxidation number of 6?

A magnesium chromate

B potassium chlorate

C barium hydrogen phosphate

D sodium hexacyanoferrate (III)

E magnesium ammonium phosphate hexahydrate

23 If an s-block element reacts with water to form a hydroxide and liberate hydrogen then

A an atom of the element contains less than 18 electrons

B $2H_2O(l) \rightarrow 2OH^-(aq) + H_2(g); -\Delta H$

C the element has been oxidized

D 1 mole of the hydroxide contains 2 moles of OH^- ions

E the hydroxide is soluble in water

	STATEMENT 1	STATEMENT 2
24	Alums containing lithium do not exist.	The lithium ion is small and has a high charge density.

25 Molten sodium is a good conductor of electricity because

1 sodium has a low ionization energy

2 sodium is a Group 1 metal

3 the 3s electron is able to act as a charge carrier

4 all the compounds of sodium are ionic in character

5 p-BLOCK ELEMENTS Al, C, N, O, Si, P, S, F, Cl, Br, I

1 The shape of an isolated ammonia molecule is

A equilateral triangular
B tetrahedral
C bipyramidal
D octahedral
E rhombic

2 In a compound of structure $X^+ \equiv O^-$ the element X could be

A sulphur
B chlorine
C aluminium
D carbon
E nitrogen

3 Which of the following compounds has the *highest* boiling point?

A hydrogen fluoride
B hydrogen sulphide
C ammonia
D phosphine
E methane

4 All of the following apply to the oxidation of nitrogen to form nitrogen monoxide [nitrogen (I) oxide] EXCEPT

A the reaction is endothermic
B the reaction will attain an equilibrium under controlled conditions
C increase in temperature changes the equilibrium constant
D the amount of nitrogen (I) oxide produced is independent of the partial pressure of oxygen
E the change in free energy of the system is positive when the reaction proceeds from left to right

5 The element which consists of atoms with less than 9 electrons orbiting the nucleus is

A phosphorus
B sulphur
C oxygen
D chlorine
E aluminium

6 Which of the following occur when aluminium powder is added to a hot aqueous solution containing sodium nitrate and sodium hydroxide?

A ammonia is evolved
B aluminium nitrate is formed in solution
C a complex ion, AlO_2^{2-}, is formed
D the nitrate ion is reduced to the nitrite
E $Al(OH_2)_6(aq) + H_2O(l) \rightarrow Al(OH_2)_5.(OH)^{2+}(aq) + H_3O^+(aq)$

7 In which of the following compounds does a p-block element have an oxidation number of 1?

A aluminium oxide
B hydrogen fluoride
C sodium hypochlorite
D phosphorus III chloride
E sulphur dioxide

8 All of the following apply to (white) phosphorus and sulphur EXCEPT that they

A are poor conductors of electricity
B are soluble in warm benzene
C combine with chlorine to form covalent chlorides
D combine exothermically with warm sodium
E yield a sulphide with hot, concentrated, aqueous sodium hydroxide.

9 Which of the following occur when dilute aqueous solutions of disodium hydrogen orthophosphate and silver nitrate are mixed?

A free orthophosphoric acid is formed in solution
B black amorphous silver oxide is precipitated
C a yellow precipitate of silver orthophosphate is formed
D heat of neutralization of the acid salt is evolved
E the pH of the solution increases

10 In order to prepare sodium thiosulphate

A sulphur is boiled with aqueous sodium sulphate until the pH of the solution is approximately 7
B sulphur dioxide is passed into hot concentrated aqueous sodium hydroxide solution until the pH = 6
C sodium hydrogen sulphate is thermally decomposed
D a saturated aqueous solution of sodium hydrogen sulphite is cooled slowly to 273 K
E electrolyse ice-cold 50% sulphuric acid with platinum electrodes

11 Which of the following applies to nitrogen?

1 marked increase in reactivity above 1 000 K
2 forms a simple molecule of structure $N \equiv N$
3 more soluble in water than oxygen at a given temperature
4 forms a hydride which burns in oxygen to yield nitrogen

12 Carbon will produce

1 carbon dioxide on reaction with concentrated nitric acid
2 carbon disulphide on reaction with sulphur
3 sodium on reaction with fused sodium hydroxide
4 calcium acetylide on reaction with calcium oxide

13 Which of the given comparisons between the named halides are correct?

	Hydrogen chloride	*Hydrogen bromide*
1	partly dissociated at 800 K	undissociated at 800 K
2	unionized in toluene	ionized in toluene
3	white precipitate of a chloride on addition to aqueous lead acetate	red-brown precipitate of a bromide on addition to aqueous lead acetate
4	oxidized only with difficulty by atmospheric oxygen	readily oxidized by atmospheric oxygen

14 An oxy-acid of phosphorus may be prepared by

1 heating red phosphorus with hot concentrated nitric acid
2 heating orthophosphoric acid at a temperature below 300 K
3 dissolving phosphorus pentoxide in water
4 passing steam over hot calcium phosphate

15 By comparison with other halogens fluorine has

1 the highest melting point
2 the lowest electronegativity
3 the smallest atomic radius
4 the greatest ability to displace oxygen from water

	STATEMENT 1	STATEMENT 2
16	Both caesium and iodine attain the xenon configuration in caesium ioxide.	Caesium iodide is covalent in character.
17	Aqueous solutions of hypochlorites are alkaline to litmus.	$ClO^-(aq) + 2H^+(aq) + 2e^- \rightleftharpoons Cl^-(aq) + H_2O(l)$
18	A possible structure for the nitrate ion is 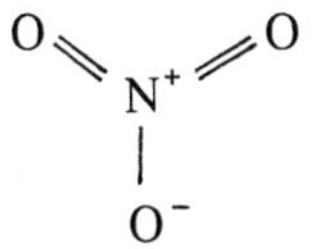	The given structure allows each atom present to attain an inert configuration.
19	Aluminium will not combine directly with carbon even on strong heating.	Aluminium and carbon are in the same Period of the Periodic Table.
20	Nitrous acid is readily oxidized to nitric acid.	$NO_2^-(aq) + H_2O(l) \rightleftharpoons NO_3^-(aq) + 2H^+(aq)$

21 The reactants and drying agents necessary to obtain a good yield of dry hydrogen bromide are

	Reactants	*Drying agents*
A	potassium bromide, concentrated sulphuric acid	concentrated sulphuric acid
B	potassium bromide, concentrated sodium hydroxide	concentrated sulphuric acid
C	potassium bromide, orthophosphoric acid	calcium bromide
D	sodium hypochlorite, potassium bromide, dilute hydrochloric acid	calcium bromide
E	hydrogen sulphide, bromine water	phosphorus V oxide

22 Which of the following aqueous solutions will decolourize aqueous iodine and give a white precipitate with aqueous lead acetate?

A hydrogen sulphide
B chlorine
C sodium sulphite
D potassium nitrate
E sodium hypochlorite

	STATEMENT 1	STATEMENT 2
23	In nitrogen III chloride nitrogen does not attain an inert configuration.	Nitrogen III chloride is a dangerously explosive yellow oil.

24 Hydrogen sulphide will react chemically as shown in

1 $2MnO_4^- (aq) + 6H^+ (aq) + 5H_2S(g) \rightarrow 2Mn^{2+} (aq) + 8H_2O(l) + 5S(s)$

2 $2Cr_2O_7^{2-} (aq) + 8H^+ (aq) + 3H_2S(g) \rightarrow 2Cr^{3+} (aq) + 7H_2O(l) + 3S(s)$

3 $2Fe^{3+} (aq) + H_2S(g) + 2Fe^{2+} (aq) \rightarrow 2H^+ (aq) + S(s)$

4 $2OH^- (aq) + H_2S(g) \rightarrow S^{2-} (aq) + 2H_2O(l)$

25 Which of the following are applicable to an aqueous solution of aluminium sulphate?

1 has a pH > 7
2 will liberate carbon dioxide from sodium carbonate
3 on thermal decomposition will give the oxide
4 forms a crystal isomorphous with potassium sulphate after concentration and cooling

6 d-BLOCK ELEMENTS, SOME TRANSITION ELEMENTS

1 Which of the following ions is blue when hydrated?

A Ti^{3+}
B Cr^{3+}
C Mn^{3+}
D Fe^{3+}
E Co^{3+}

2 The formula of the complex cyanide which is the most stable to oxidation is

A $K_3[V(CN)_6]$
B $K_3[Cr(CN)_6]$
C $K_3[Mn(CN)_6]$
D $K_3[Fe(CN)_6]$
E $K_3[Co(CN)_6]$

3 Which of the following ions will have the largest radius?

A V^{2+}
B Ti^{4+}
C Mn^{2+}
D Cr^{3+}
E Fe^{3+}

4 Which of the following elements has the electronic configuration

(argon core) $3d^6\ 4s^2$

A iron
B cobalt
C manganese
D nickel
E vanadium

5 The maximum oxidation number of chromium is 6 because

A chromium forms an ion of formula CrO_4^{2-}
B potassium dichromate is formed when potassium chromate is oxidized
C $\frac{1}{6}$ of a mole of potassium chromate is equivalent to 1 Faraday of charge
D the seventh electron would have to be removed from the argon core
E all d-block elements can have an oxidation number of 6

6 The rate of oxidation of the iodide ion according to the reaction

$$2I^-(aq) + S_2O_8^{2-}(aq) \rightarrow I_2(s) + 2SO_4^{2-}(aq)$$

is markedly increased by the addition of any of the following ions EXCEPT

A Mn(VII)
B Ni(II)
C Fe(II)
D Fe(III)
E Cu(II)

7 When chlorine was passed through a strongly alkaline suspension of a hydroxide the colour changed from green to yellow.
The hydroxide contained ions of formula

A Fe^{3+}
B Cr^{3+}
C Mn^{2+}
D Co^{2+}
E Ni^{2+}

8 Which of the following is a correct statement of what will occur when aqueous solutions containing elements with the given oxidation numbers are mixed?

A	V(V)	+	Br(-1)	→	V(IV)	+	Br(O)
B	V(IV)	+	I(O)	→	V(V)	+	I(-1)
C	V(IV)	+	V(II)	→	V(III)		
D	V(IV)	+	Fe(III)	→	V(V)	+	Fe(II)
E	V(III)	+	Fe(II)	→	V(IV)	+	Fe(III)

9 Which of the following reagents are normally used in the laboratory preparation of the compound of formula $K^+.Cr^{3+}.2SO_4{}^{2-}.12H_2O$?

A potassium dichromate, sulphuric acid, ethanol
B potassium sulphate, chromium (III) sulphate, sulphuric acid
C chromium (III) sulphate, potassium hydroxide, sulphur dioxide
D potassium chromate, sulphuric acid
E potassium hydroxide, chromium (III) hydroxide, sulphuric acid

10 Which of the following complex ions can exhibit optical isomerism?

A $[Cu(NH_3)_4]^{2+}$
B $[Co(CN)_6]^{3-}$
C $[Cr(\text{ethanediamine})_2Cl_2]^+$
D $[Fe(SCN)]^{2+}$
E $[Co(NO_2)_6]^{3-}$

11 When potassium permanganate is added to a saturated aqueous solution of potassium hydroxide

1 a gas is evolved
2 the gas is oxygen
3 the oxidation number of oxygen changes from (-II) to (0)
4 2 moles of oxygen are evolved per mole of permanganate reacting

12 Which of the following apply to an aqueous solution of potassium hexacyanoferrate (II)?

1 the solution is colourless
2 gives a pale blue colouration with an aqueous solution of an iron (III) salt
3 is a solution containing a complex ion which has 6 ligands
4 will evolve hydrogen when warmed with a dilute mineral acid

13 Which of the following occur when hydrogen peroxide is added to an acidified aqueous solution of a chromate?

1 a deep blue colour is formed initially
2 after some time the solution turns yellow
3 oxygen is evolved
4 the electronic configuration of chromium in the resulting solution is (argon core) $3d^6$

14 Hydrated copper (II) sulphate is blue in colour because
1 4 molecules of water are attached to 1 atom of copper
2 there is a change in free energy of the system when anhydrous copper (II) sulphate gains water
3 the sulphate ion is covalent in structure and forms a resonance hybrid
4 light at the red end of the visible spectrum is absorbed

15 Which of the following are evidence that titanium (IV) compounds are more stable than copper (II) compounds?
1 4 is the highest oxidation number for titanium, 2 for copper
2 copper (II) oxide can be reduced to copper by ethanol
3 copper has a larger atomic number than titanium
4 $\Delta G_{f, 298}$/kJ g-atom^{-1} of element is less for titanium (IV) chloride than for copper (II) chloride

	STATEMENT 1	STATEMENT 2
16	$CuCl_4^{2-}$ ions are yellow in colour in aqueous solution.	Addition of concentrated hydrochloric acid to aqueous copper (II) sulphate produces a yellow colour.
17	To produce a solution containing the highest concentration of the complex Ni(II)-EDTA ion equimolar solutions of a Ni(II) salt and EDTA (disodium salt) should be mixed.	The pH of a solution containing a Ni(II) salt is raised when EDTA (disodium salt) is added as 2 moles of H^+ ion are liberated per mole of EDTA salt added.
18	When zinc dust is added to an aqueous solution containing V(V) ions the colour changes from yellow to blue.	Zinc has been oxidized and the oxidation number of vanadium has decreased by 3 units.
19	Copper attains an inert configuration in the ion $Cu(NH_3)_4^{2+}$.	Each nitrogen atom donates 2 electrons to copper in the tetramminecopper (II) ion.

	STATEMENT 1	STATEMENT 2
20	Nickel is used to catalyse the hydrogenation of unsaturated hydrocarbons.	Nickel allows the hydrogenation to proceed at a lower activation energy.

Questions 21 to 25 refer to the list of colours given below. Select the response which you think answers the question correctly.

A violet **C** red **E** pink

B green **D** blue

Which of the above is the colour

21 of the hydrated Mn^{2+} ion?

22 produced when excess sulphur dioxide is passed through ammonium metavanadate solution?

23 observed when sodium 2-hydroxybenzoate is added to an aqueous solution of the tetrammine copper (II) ion?

24 observed when a few drops of 0·1 M potassium permanganate is added to a concentrated acidified solution of hydrated manganese (II) sulphate?

25 observed when either zinc or tin dust is added to an acidified solution of titanium (IV) chloride?

7 OXIDATION AND REDUCTION, HALF CELLS

1 Which of the following applies to the reaction given below?
$MnO_4^-(aq) + 8H^+(aq) + 5Fe^{2+}(aq)$
$\rightleftharpoons 5Fe^{3+}(aq) + Mn^{2+}(aq) + 4H_2O(l)$

A hydrogen ions catalyse the reaction
B 1 mole of oxidant is equivalent to 0·2 moles of reductant
C 5 Faradays of charge are transferred per mole of permanganate ion
D the reaction will not proceed at room temperature
E a suitable source of H^+ ions is dilute hydrochloric acid

2 If a reaction occurs between the given pairs of reactants in which reaction would 96 500 C be transferred when 0·25 of a mole of reductant is oxidized?

A acidified potassium dichromate, iron (II) ammonium sulphate
B acidified potassium permanganate, sodium oxalate
C sodium thiosulphate, potassium iodide
D acidified iron (II) sulphate, hydrogen peroxide
E potassium tetroxalate, acidified potassium permanganate

3 For the reaction
$Zn(s) + Cu^{2+}(aq) \rightleftharpoons Zn^{2+}(aq) + Cu(s)$

A the system is a voltaic cell and the emf = lg $[Zn^{2+}(aq)]$ eqm
B at equilibrium the ratio $Zn^{2+}(aq)/Cu^{2+}(aq)$ is a constant
C the rate of reaction is independent of the surface area of the reactants
D the reaction is only reversible when $Zn^{2+}(aq) = Cu^{2+}(aq)$
E no energy is lost during the reaction as the same charge is transferred for the forward and reverse reactions

4 What is the change in oxidation number of chlorine when sodium chlorate is reduced to sodium chloride?

A 0 **C** 5 **E** 8
B 1 **D** 6

5 What is the difference between the ionization energy of copper and the energy required to form 1 mole of the Cu^{2+}(aq) ion from the metal at 298 K, in kJ per mole?

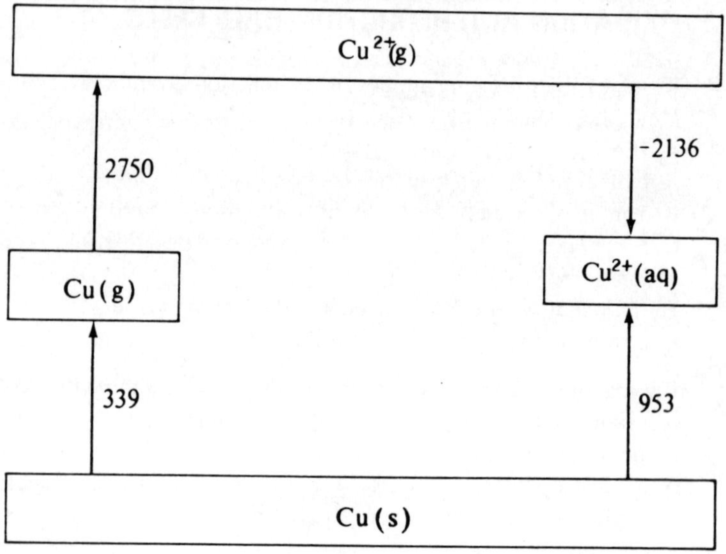

A 614
B 1 173
C 1 897
D 3 089
E 4 886

6 Which of the following apply to the cell shown in the diagram when the gases are flowing?

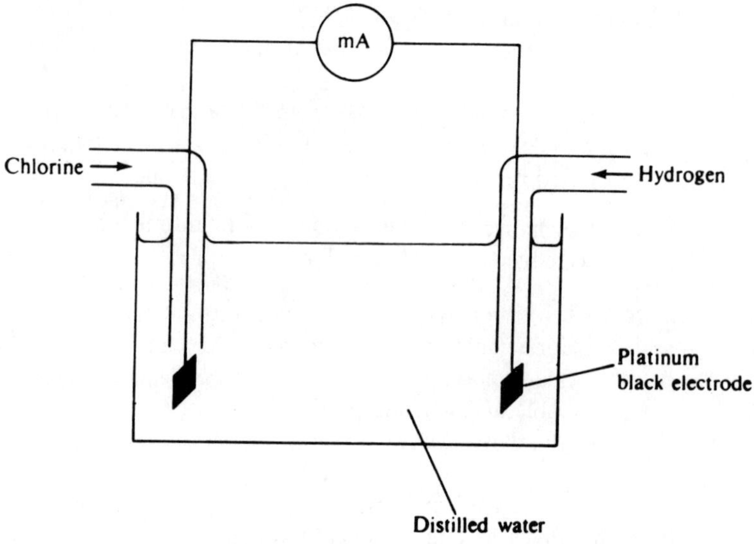

A negative charge will tend to flow in a clockwise direction through the external circuit
B platinum black increases the solubility of the gases in the electrolyte
C the electrodes are inert and reversible
D $Pt | H_2(aq) \,\vdots\, Cl_2(aq) | Pt$
E work is obtained from the system and its free energy increases

7 In which of the following reactions is 1 mole of oxidant equivalent to 6 moles of reductant and the oxidation number of the effective element present in the reductant increased by unity?
A $IO_3^-(aq) + 5I^-(aq) + 6H^+(aq) \rightarrow 3H_2O(l) + 3I_2(s)$
B $I_2(s) + 2S_2O_3^{2-}(aq) \rightarrow 2I^-(aq) + S_4O_6^{2-}(aq)$
C $2CrO_4^-(aq) + 2H^+(aq) \rightarrow Cr_2O_7^{2-}(aq) + H_2O(l)$
D $Sn^{2+}(aq) + 2Fe^{3+}(aq) \rightarrow Sn^{4+}(aq) + 2Fe^{2+}(aq)$
E $SO_3^{2-}(aq) + I_2(s) + 2HCO_3(aq) \rightarrow SO_4^{2-}(aq) + H_2O(l) + CO_2(g)$

8 Which of the following conclusions may be drawn from the graph of emf against lg [ion] for the simple cell consisting of chromium and copper electrodes?

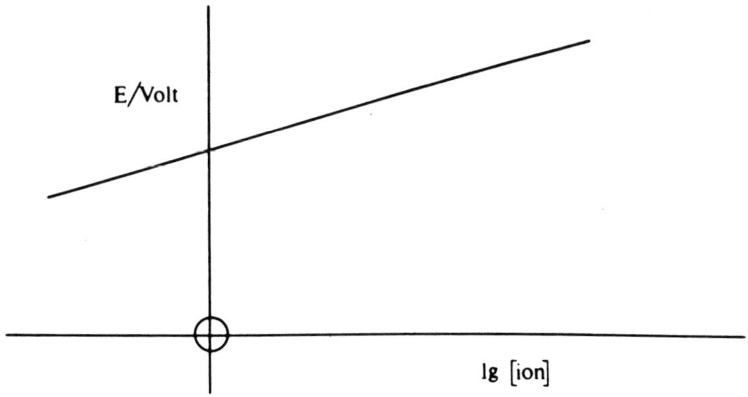

A the $[Cu^{2+}]$ ion would have to be changed at the same rate as $[Cr^{3+}]$ for a constant flow of charge to be maintained
B the standard redox potential of one of the electrodes is given by the value of E when lg [ion] = 0
C the conductivity of the electrolyte increases as the dilution increases
D the emf is dependant upon the type of electrodes used
E $E_{cell} = E^{\ominus}_{Cr} + E^{\ominus}_{Cu}$ at all concentrations

9 All of the following are correct statements EXCEPT

A an oxidant takes up electrons in ionic oxidation

B covalent reduction is the removal of a more electronegative element from a compound

C a molecule can be oxidized by electron loss

D the cathode of a simple cell is the electrode at which reduction occurs

E Avogadro's constant is equivalent to 1 Faraday of negative charge

10 In which of the following reactions is the acid the oxidant?

A $KI(s) + H_3PO_4(l) \rightarrow HI(g) + KH_2PO_4(s)$

B $3I_2(s) + 10HNO_3(l) \rightarrow 6HIO_3(l) + 10NO(g) + 2H_2O(l)$

C $2MnO_2(s) + 2H_2SO_4(l) \rightarrow 2MnSO_4(s) + 2H_2O(g) + O_2(g)$

D $4CrO_3(s) + 2H_2SO_4(l) \rightarrow 2Cr_2(SO_4)_3(s) + 6H_2O(l) + 3O_2(g)$

E $4HClO_4(l) + P_4O_{10}(s) \rightarrow 2Cl_2O_7(g) + 4HPO_3(s)$

11 Which of the following will catalyse the thermal decomposition of potassium chlorate?

1 iron (III) oxide

2 cobalt (II) oxide

3 nickel (II) oxide

4 copper (I) oxide

12 Which of the following apply to the Nernst equation in the form

$$E = E^{\ominus} + \frac{RT}{zF} \lg \frac{[\text{oxidized form}]}{[\text{reduced form}]}$$

for the equilibrium $V^{5+} + 3e^- \rightleftharpoons V^{2+}$?

$V^{5+} + 3e^- \rightleftharpoons V^{2+}$?

1 $z = 2$

2 T is the thermodynamic temperature K

3 R is the gas constant ($J\,K^{-1}$)

4 lg [oxidized form] is the logarithm of the V^{5+} ion concentration ($mol\ dm^{-3}$)

13 If a cell is constructed from the following electrode systems

$Sn^{2+}(aq) \mid Sn(s)$; $E = -0{\cdot}14$ volt

$Br_2(aq), 2Br^-(aq) \mid Pt$; $E = +1{\cdot}09$ volt

1 the $Br_2(aq), 2Br^-(aq) \mid Pt$ system will be the positive pole

2 $Sn^{2+}(aq)$ is a better oxidant than $Br_2(l)$

3 the emf of the system will be 1·23 volt

4 in the working cell the reaction represented by the lower electrode system goes from right to left

14 Which of the following can be deduced from the redox diagram given below?

0·90 volt $VO_2^+(aq) + 2H^+(aq) + e^- \rightleftharpoons H_2O(l) + VO^{2+}(aq)$

1·09 volt $Br_2(l) + 2e \rightleftharpoons 2Br^-(aq)$

1·33 volt $Cr_2O_7^{2-}(aq) + 14H^+(aq) + 6e^- \rightleftharpoons 2Cr^{3+}(aq) + 7H_2O(l)$

1 potassium dichromate will readily oxidize the bromide ion

2 chromium and vanadium have the same principal oxidation numbers

3 $[VO_2^+(aq) + 2H^+(aq)], [VO^{2+}(aq) + H_2O(l)] \mid Pt; E = 0{\cdot}90$ volt

4 the vanadium ion VO_2^+ will oxidize the Br ion more easily than the $Cr_2O_7^{2-}$ ion

15 Which of the following apply to the reaction

$$S_2O_6^{2-}(aq) + 4H^+(aq) + 2e^- \rightleftharpoons 2H_2O(l) + 2SO_2(g)$$

1 the equilibrium constant, $K = \dfrac{[SO_2(aq)]^2 \text{ eqm}}{[S_2O_6^{2-}(aq)] \text{ eqm}}$

2 the reaction proceeds to completion when the pH = 4

3 1 mole of $S_2O_6^{2-}$ ions liberates 22·4 dm^3 of sulphur dioxide at 298 K

4 sulphur exhibits the oxidation numbers 5 and 4

	STATEMENT 1	STATEMENT 2
16	The addition of zinc dust to ammonium metavanadate produces a change in colour from yellow to violet.	$VO^{2+}(aq) + 2H^+(aq) + 2e^- \rightleftharpoons H_2O(l) + V^{3+}(aq)$
17	Only compounds containing atoms which have an incomplete penultimate electron shell are strong oxidizing agents.	Hydrogen peroxide can be reduced by acidified potassium permanganate.
18	$I_3^-(aq) + 2S_2O_3^{2-}(aq) \rightleftharpoons 3I^-(aq) + S_4O_6^{2-}(aq)$	$I_2(s) + 2e^- \rightleftharpoons 2I^-(aq)$; $E = 0{\cdot}54$ V $S_4O_6^{2-}(aq) + 2e^- \rightleftharpoons 2S_2O_3^{2-}(aq)$; $E = 0{\cdot}08$ V

19 When zinc powder is added to a saturated aqueous solution of copper (II) sulphate the system increases in temperature because

A zinc powder has a large surface area

B the reaction is a redox process

C zinc requires less energy to ionize than copper

D $\Delta H^{\ominus}_{298K}$ Zn(s) = 0

E 2 Faradays of charge are transferred per g-atom of reactants

	STATEMENT 1	STATEMENT 2
20	The cell represented by $Mg(s) \mid Mg^{2+}(aq) \vdots Fe^{2+}(aq) \| Fe(s)$ is unsuitable for use as a source of electrical energy	The standard electrode potentials for the electrodes $Mg^{2+}(aq) \mid Mg(s)$ and $Fe^{2+}(aq) \mid Fe(s)$ are both less than zero.
21	The half cell diagram for potassium permanganate in acid solution can be written as $[MnO_4^-(aq) + 8H^+(aq)]$, $[Mn^{2+}(aq) + 4H_2O(l)] \mid Pt$.	Water is a product of the oxidation of potassium permanganate and potassium permanganate is an oxidant in the presence of hydrogen ions.
22	A low resistance voltmeter is suitable for determining the emf of a voltaic cell.	A voltaic cell generates a small current.

23 If a platinum wire is dipped into a solution containing $Fe^{2+}(aq)$ and $Fe^{3+}(aq)$ is may be concluded that

1 the electrode system is $Pt \mid Fe^{2+}(aq), Fe^{3+}(aq)$
2 the reaction at the electrode is $Fe^{2+}(aq) \rightleftharpoons Fe^{3+}(aq) + e^-$
3 the electrode may either donate or accept electrons
4 platinum is inert with respect to the aqueous ions

24 All of the redox reactions given below are correct EXCEPT

A $Fe^{2+}(aq) + Fe(CN)_6^{3-}(aq) \rightleftharpoons Fe^{3+}(aq) + Fe(CN)_6^{4-}(aq)$
B $2Fe^{2+}(aq) + 2H^+(aq) + H_2O_2(aq) \rightleftharpoons 2Fe^{3+}(aq) + 2H_2O(l)$
C $2Fe^{2+}(aq) + Br_2(l) \rightleftharpoons 2Fe^{3+}(aq) + 2Br^-(aq)$
D $6Fe^{2+}(aq) + 8H^+(aq) + 2NO_3^-(aq)$
$\rightleftharpoons 6Fe^{3+}(aq) + 4H_2O(l) + 2NO(g)$
E $2Fe^{2+}(aq) + SO_2(g) + H_2O(l) \rightleftharpoons$
$\rightleftharpoons 2Fe^{3+}(aq) + SO_3^{2-}(aq) + 2H^+(aq)$

25 In which of the following compounds does an element exhibit an oxidation number of −3?

A nitric acid
B ammonium chloride
C chromium(III) sulphate
D potassium iodate
E hydrogen peroxide

8 ENTHALPY, ENTROPY AND BOND ENERGIES

Each of Questions 1 to 5 refer to the energy level diagram for the formation of silver chloride. Select the response from the diagram which you think answers the question correctly.

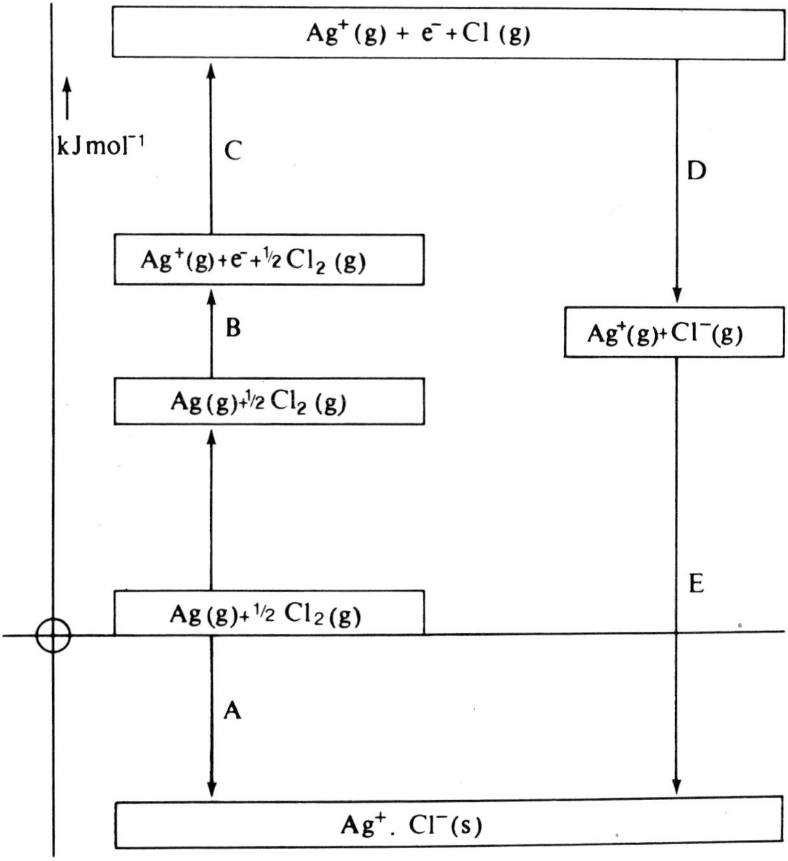

Which of the above

1 is a measure of the electron affinity of chlorine?

2 may be represented as $\Delta H_{f, 298}^{\ominus}$ [$Ag^+ Cl^-(s)$] ?

3 is the ionization energy of silver?

4 is the heat of atomization of chlorine?

5 is the lattice energy of crystalline silver chloride?

6 Use the data on the graph to determine which compound has a standard heat of formation per mole of -395·4 kJ.

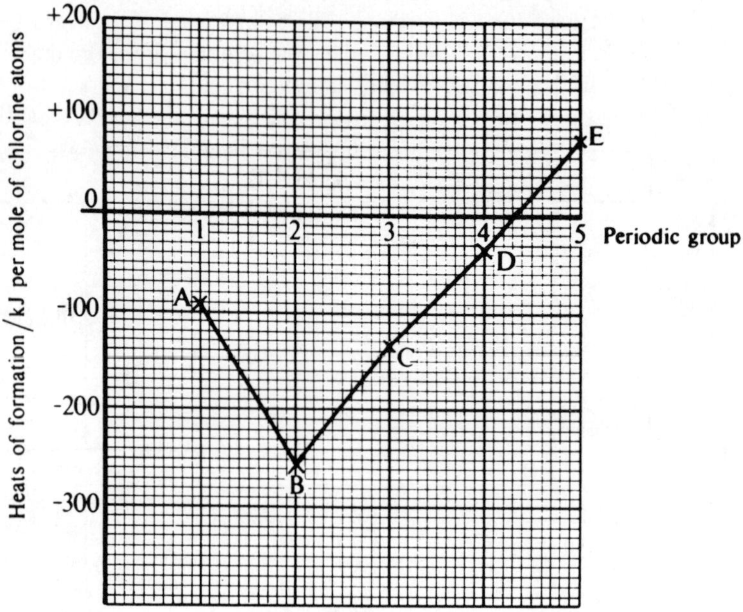

7 The standard heat of formation can be obtained by a single stage chemical reaction for

A lead (II) nitrate
B manganese (IV) oxide
C ethane
D potassium chloride
E dinitrogen oxide

8 Which of the following is chemically and thermodynamically correct?

A $C\ (diamond) + O_2\ (g) \rightarrow CO_2\ (g);\ +\Delta H^{\ominus}_{f,\ 298}$

B $H_2SO_4\ (aq)\ (2M) + 4\ NaOH(s)$
$\rightarrow Na_2SO_4(aq)\ (4M) + H_2O\ (aq);\ -\Delta H^{\ominus}_{298}$

C $CS_2\ (l) + 3O_2\ (g) \rightarrow CO_2\ (g) + 2SO_2\ (g);\ -\Delta H^{\ominus}_{combustion,\ 298}$

D $2Cl^-\ (g) \rightarrow Cl_2\ (g) + 2e^- - \Delta H^{\ominus}_{298}$

E $2CH_3OH(l) + O_2\ (g) \rightarrow 2CO(g) + 2H_2O(l);\ -\Delta H^{\ominus}_{298}$

9 In order to determine the bond energy of one C–H bond in a saturated hydrocarbon the information required is

A the heat of atomization and the molecular formula of the compound

B the heat of formation and the empirical formula of the compound

C the heat of combustion and the number of moles of water produced per mole of hydrocarbon

D the heat of formation of the hydrogen molecule and the number of hydrogen atoms per mole of compound

E the heat of formation of carbon dioxide and the heat of combustion of the compound

10 Why should the heat of combustion of ethanal and epoxyethane be different if they both have the same molecular formula?

A ethanal is a liquid at room temperature

B the bond energies between the constituent atoms in ethanal and epoxyethane are different

C epoxyethane is a result of the partial oxidation of ethene

D the methyl group in ethanal is stable to oxidation

E the bond length between carbon and hydrogen is different for different alkanes

	STATEMENT 1	STATEMENT 2
11	A system will increase in internal energy if the entropy of the system is increased.	The entropy of a system is a function of the number of different ways in which the total energy of the system can be distributed.
12	$\Delta H^{\ominus}_{298}$ [P_4(s)] = 0	The absolute enthalpy of an element can not be determined, only changes in enthalpy can be calculated.
13	The Law of Conservation of Energy only applies to a chemical reaction if no energy is lost to the surroundings.	The standard heat of formation of a compound depends upon the manner in which the compound is formed.
14	Hydrogen iodide will be less thermally stable than hydrogen chloride.	Energy is absorbed from the surroundings when hydrogen iodide is formed and lost to the surroundings when hydrogen chloride is formed from its elements.

	STATEMENT 1	STATEMENT 2
15	The reaction $Pb^{2+}(aq) + 2Cl^-(aq) \rightarrow PbCl_2(s)$ is exothermic.	The change $Pb(g) \rightarrow Pb^{2+}(g) + 2e^-$ is endothermic.
16	When a mass of gas is heated at constant volume the translation and rotational energy of the molecules are increased, the vibrational energy remains constant.	In most chemical reactions the increase in internal energy is small and the change in enthalpy during a chemical reaction between compounds is a measure of the bonding energy between the atoms.
17	The enthalpy of combustion of a compound can be determined if the bond energy for each of the component bonds in 1 mole of the compound is known.	The heat of combustion of butadiene is the sum of the bond energies of two C = C and six C-H bonds.

18 In order to determine the enthalpy change in the reaction

$$Hg^{2+}(aq) + Cu(s) \rightarrow Hg(l) + Cu^{2+}(aq)$$

it is necessary to know the enthalpy change in

1 $Zn(s) + Cu^{2+}(aq) \rightarrow Zn^{2+}(aq) + Cu(s)$
2 $HgCl_2(s) = aq \rightarrow HgCl_2(aq)$
3 $Zn(s) + Hg^{2+}(aq) \rightarrow Zn^{2}(aq) + Hg(l)$
4 $HgCl_2(aq) \rightarrow Hg^{2+}(aq) + 2Cl^-(aq)$

19 Sodium is more chemically reactive than aluminium because

1 $\Delta H^{\ominus}_{at,298}$ [Na(g)] $< \Delta H^{\ominus}_{at,298}$ [Al(g)]
2 $\Delta H^{\ominus}_{f,298}$ [NaCl(s)] $< \Delta H^{\ominus}_{f,298}$ [$AlCl_3$(g)]
3 the molar enthalpy change at melting point is greater for aluminium than for sodium
4 the 1st ionization energy of sodium is less than that of aluminium

20 Which of the following support the theory of Periodicity of the elements?

1 the heat of neutralization of molar aqueous solutions of hydrogen chloride and hydrogen bromide with aqueous sodium hydroxide are approximately the same
2 the lattice energy of lithium fluoride is less than that of lithium iodide

3 the heat of atomization of graphite is not the same as that of diamond

4 the molar heat capacities of the chlorides of magnesium, calcium and strontium are approximately the same

21 If 1 mole of sodium is added to an excess of water in a thermally insulated container under constant volume conditions then

A 1 mole of hydrogen is produced

B the total enthalpy change is the heat of ionization of 1 mole of sodium atoms

C the energy change is an increase in internal energy

D the enthalpy of the system increases

E the rate of reaction increases as no external work is done

22 At a temperature of 368·5 K rhombic sulphur slowly changes into monoclinic sulphur, therefore

A rhombic sulphur is kinetically stable but energetically unstable

B the S-S bond energy is greater for rhombic sulphur than monoclinic sulphur

C the standard heat of formation of an allotrope is not zero

D when a change in crystalline shape occurs the enthalpy of the system increases

E there are more atoms per mole of rhombic sulphur than monoclinic sulphur

	STATEMENT 1	STATEMENT 2
23	Crystals of anhydrous calcium chloride lose energy on exposure to the atmosphere.	$CaCl_2(s) + aq$ $\rightarrow CaCl_2.6H_2O(s); -\Delta H$

24 In order to calculate the heat change for a reaction occuring in a closed calorimetric system it is necessary to know

1 the heat capacity of the calorimeter

2 the heat capacity of the contents of the calorimeter

3 the concentration of the reactants

4 the temperature change of the system

	STATEMENT 1	STATEMENT 2
25	$\Delta H^{\ominus} - T\Delta S = \Delta G^{\ominus}$	$\Delta G^{\ominus}$ varies with temperature and may be used to determine the equilibrium constant for a gaseous reaction.

9 CHEMICAL AND PHASE EQUILIBRIA, HYDROGEN BONDING, SOLUBILITY

1 The variation of vapour pressure with temperature for two homogeneous liquid mixtures, X and Y, is as shown; X and Y are formed by mixing two liquids in different proportions.

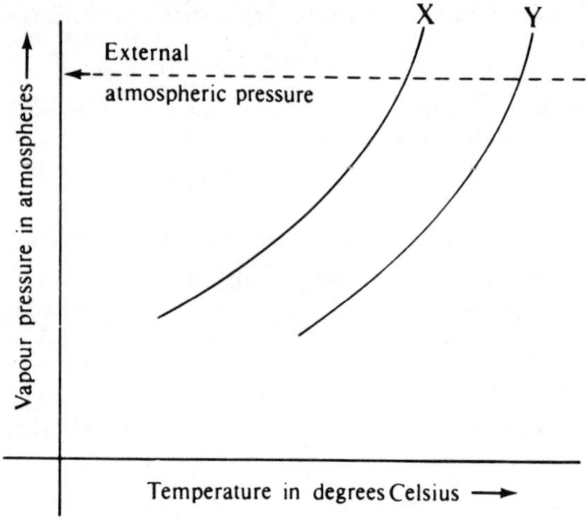

It may be deduced from the graph that

A for a mixture of two liquids the vapour pressure is the sum of the partial pressures of each component

B a mixture boils when its vapour pressure is equal to that of the atmosphere

C a decrease in vapour pressure causes an increase in boiling point

D for an ideal mixture the vapour pressure is proportional to the change in composition of the mixture

E hydrogen bonding has occurred in mixture Y

2 The energy required to cause the sublimation of 1 mole of helium is only about $\frac{1}{400}$ that required for 1 mole of ice because

A (heat of sublimation) = (heat of fusion) + (heat of vaporization)

B energy is evolved when helium freezes

C helium is an inert gas

D work has to be done to overcome hydrogen bonding when ice is sublimed

E 1 mole of ice is much heavier than 1 mole of helium

3 Which of the following apply to the reaction

$$N_2(g) + 3H_2(g) \rightleftharpoons 2NH_3(g)$$

A $K_p = \dfrac{[NH_3(g)]^2}{[N_2(g)]\ [H_2(g)]^3}$

B the partial pressure of hydrogen
$= \left[\dfrac{\text{no. of moles of hydrogen}}{\text{total no. of moles present}}\right] \times$ [(total pressure of system)]

C $[N_2(g)]$ is proportional to the volume of nitrogen present

D at low pressure the equilibrium lies well to the right

E the reaction will proceed to equilibrium only at 200 atm and 723 K

4 Which of the following apply to the reaction
$Zn(s) + Cu^{2+}(aq) \rightarrow Zn^{2+}(aq) + Cu(s)$; $\Delta H^{\ominus}_{298} = -217$ kJ

A the concentration of all pure solid phases can be considered constant

B $K_c = \dfrac{[Zn^{2+}(aq)]}{[Cu^{2+}(aq)]}$

C at equilibrium the system would be unaffected by temperature changes

D the rate at which the reaction proceeds is independent of $[Cu^{2+}(aq)]$

E the products of the reaction will be energetically stable

5 The formula of the compound with the highest boiling point is

A $CH_3CH_2CH_2CH_3$

B $CH_3CH_2CH_2OH$

C $CH_3CH_2CH_2CH_2CH_3$

D $CH_3CH_2OCH_2CH_3$

E $CH_2OHCHOHCH_2OH$

6 The increase in melting point of the halogens with increasing atomic number is due *principally* to an increase in the

A electronic charge surrounding the nucleus

B mass of a mole of the element

C metallic character of the elements

D atomic diameter

E heat of sublimation

7 The graph shows the variation in boiling point with concentration of a liquid P in a liquid Q.

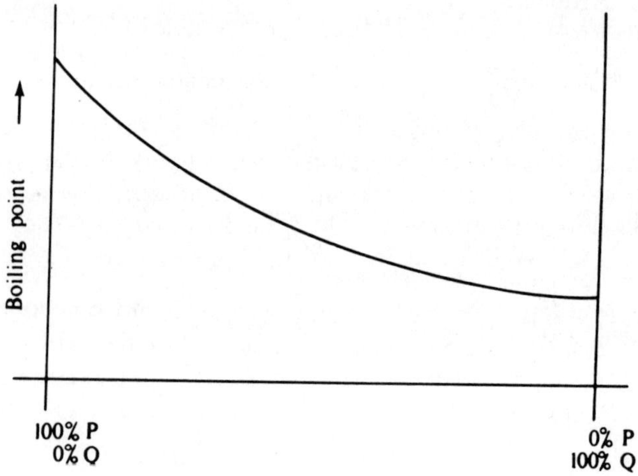

It may be concluded from the graph that

A the attraction between P molecules is equal to the attraction between P and Q molecules

B pure Q has a higher vapour pressure than pure P, at the same temperature

C the solution of P in Q is non-ideal and is a negative deviation from Raoult's Law

D an enthalpy change will occur when P and Q are mixed

E P and Q are non-ionic compounds

8 Calcium and strontium will dissolve in ammonia to yield a blue solution because

A the copper (II) ion forms a blue complex with ammonia

B all Group 2 elements are soluble in ammonia

C solvation of the metal atoms occurs to form a complex of type $Ca(NH_3)_6^{2+}$

D the metals do not displace hydrogen from ammonia

E the emission spectra for the Alkaline Earth metals is red, blue light is absorbed

	STATEMENT 1	STATEMENT 2
9	Iodine is soluble in benzene.	The attractive force between iodine molecules is similar to that between iodine and benzene molecules at room temperature.

	STATEMENT 1	STATEMENT 2
10	No temperature change occurs when 1 mole of calcium chloride is dissolved in 1 dm^3 of water at room temperature	The lattice energy of the calcium chloride crystal is small and very little hydration of the ions occurs.
11	An equimolecular mixture of tetrachloromethane and acetone has a boiling point greater than that of either constituent.	A fall in temperature occurs when equimolecular proportions of tetrachloromethane and acetone are mixed.
12	Water is the only liquid that has a measurable surface tension at room temperature.	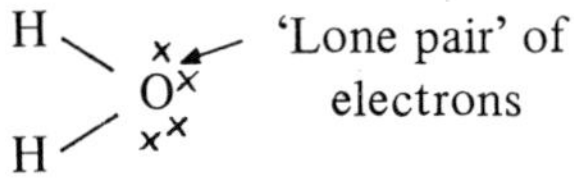
13	The partial pressure of each component of a mixture of liquids is proportional to the mole fraction of each component present in the mixture.	The total pressure of a system is equal to the sum of the partial pressures of each of the components of the system.
14	Nylon does not absorb moisture as readily as cotton.	Cotton contains many hydroxyl groups capable of undergoing hydrogen bonding with water.

15 Which of the following occurs when phosphorus (V) chloride is placed in an evacuated tube and maintained at a high constant temperature?

1 thermal dissociation
2 two phases, at equilibrium
3 a stable state of equilibrium
4 the maximum number of molecules are present at equilibrium

16 Which of the following apply to the equilibrium $Hg(l) \rightleftharpoons Hg(g)$

1 [Hg(l)] is proportional to [Hg(g)]
2 the equilibrium is independent of the initial Hg(l)
3 the equilibrium is independent of pressure
4 at equilibrium the vapour pressure of mercury is a minimum

17 When a solute is distributed between two solvents, P and Q, the ratio $\frac{[\text{solute in P}]}{[\text{solute in Q}]}$ is a constant if the

1 solvents are immiscible
2 solute is soluble in P and Q
3 temperature is maintained constant
4 solute is ionized

18 Addition of excess water to the system
$BiCl_3(aq) + H_2O(l) \rightleftharpoons BiOCl(s) + 2HCl(aq)$
will, at equilibrium
1 displace the equilibrium to the left
2 decrease the rate of the forward reaction
3 increase the solubility of bismuth (III) chloride
4 produce an increase in concentration of bismuth oxychloride

19 Acetone is a liquid at room temperature, a fine stream of which may be deflected by an electric field because
1 acetone molecules are polar
2 acetone contains two methyl groups
3 interaction between acetone molecules occurs due to dipole attraction
4 acetone molecules do not form hydrogen bonds

20 On mixing equal volumes of trichloromethane and acetone in an insulated container a fall in temperature was observed because
1 trichloromethane is more dense than acetone
2 acetone is a reducing agent
3 the compounds reacted chemically
4 trichloromethane forms hydrogen bonds with acetone

Questions 21 to 25 refer to the graphs opposite. Select the response which you think is the correct answer to the question.

Which of the graphs represent the

21 energy change during a reaction in which an unstable intermediate product is formed?

22 change in vapour pressure with increased concentration of a non-volatile solute in a volatile solvent?

23 volume change which occurs when hydrogen bonds in water are formed and then broken between 273 K and 281 K?

24 change in solubility of a hydrated salt in water if the salt loses all its water of crystallization at a particular temperature?

25 overall energy changes in an exothermic reaction?

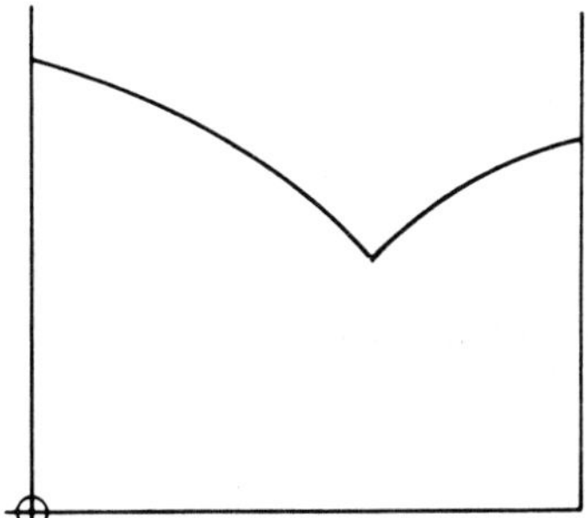

A

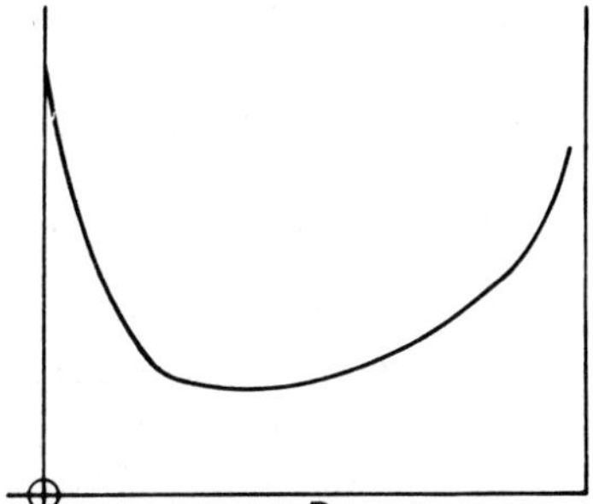

B

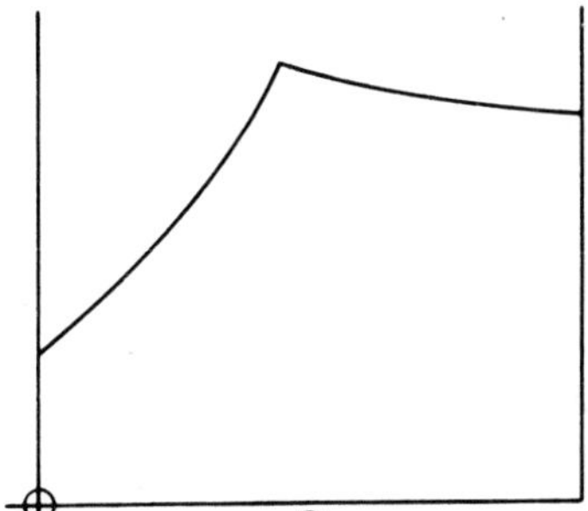

C

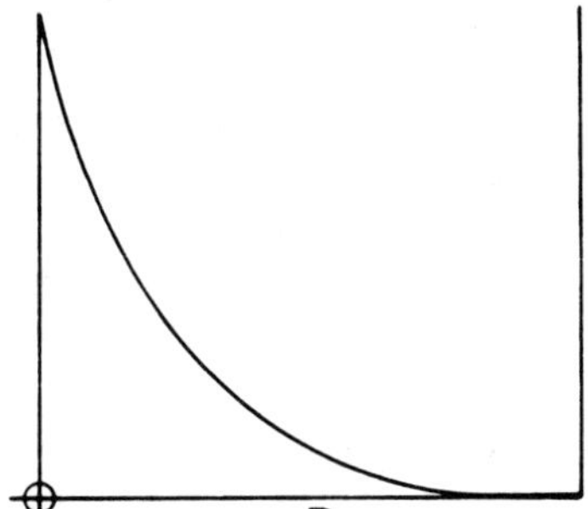

D

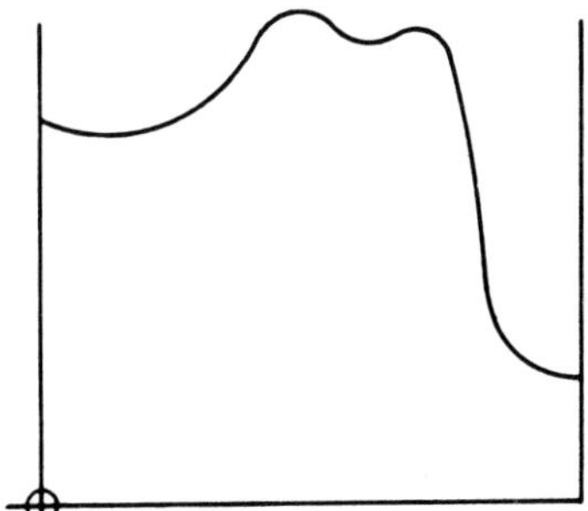

E

10 K_w, K_c, K_a, K_b, K_{sp}, pH, pK_a, k

1 Iron (III) hydroxide will be precipitated when aqueous ammonia is added to a neutral solution of an iron (III) salt because

A $NH_3(aq) + H_2O(aq) \rightleftharpoons NH_4^+(aq) + OH^-(aq)$
B the solubility product of iron (III) hydroxide is exceeded
C iron (III) hydroxide is a hydrated oxide
D the solubility product of iron (II) hydroxide > solubility product of iron (III) hydroxide
E the pH of a neutral solution = 7

2 The $[H^+(aq)]_{eqm}$ for a weak acid, in aqueous solution, is given by

A $K_a \times \dfrac{[H^+A^-]_{eqm}}{[A^-]_{eqm}}$

B $\lg(K_a)$

C $\lg \dfrac{[H^+A^-]_{eqm}}{[A^-]_{eqm}}$

D $\dfrac{K_a}{[A^-]_{eqm}}$

E $\lg(K_a) + \lg \left[\dfrac{[H^+A^-]_{eqm}}{[A^-]_{eqm}}\right]$

3 For the reaction at constant temperature of

$\frac{1}{2} N_2(g) + \frac{3}{2} H_2(g) \rightleftharpoons NH_3(g)$

the value $K_p = \dfrac{(p_{NH_3})_{eqm}}{(p_{N_2})^{\frac{1}{2}}_{eqm}\,(p_{H_2})^{\frac{3}{2}}_{eqm}}$ because

A the ratio of volumes of reactants to products is 2:1
B the equilibrium constant is a function of the concentration of the reactants
C the ratio of the rate of the forward reaction to the rate of the reverse reaction is a constant
D the concentration of a gas is proportional to its pressure at constant temperature
E $K_p = K_c$ since the difference in the number of moles of reactants compared with products is unity

4 The equilibrium constant for the ionization of water at 298 K is given as 10^{-14} because

A the pH of pure water = 7
B 1 mole of water dissociates to give 1 mole of H^+ ions

C $[H^+(aq)]_{eqm} = [OH^-(aq)]_{eqm} = 10^{-7}$ and $[H_2O(l)]_{eqm}$ may be considered constant
D the degree of dissociation of water is temperature dependent
E H^+ ions are hydrated in aqueous solution

5 The total volume of the equilibrium mixture represented by
$C_2H_5OH(l) + CH_3COOH(l) \rightleftharpoons CH_3COO.C_2H_5(l) + H_2O(l)$
is **not** needed to determine K_c because
A $K_c = \dfrac{[CH_3COO.C_2H_5(l)]_{eqm}}{[CH_3COOH(l)]_{eqm}}$
B K_c is determined only by the number of moles of alcohol present initially
C each reactant can be considered to occupy the same volume
D the rate of reverse reaction is negligible at room temperature
E the reaction is catalysed by H^+ ions

6 Which of the following dilute aqueous solutions has a pH $>$ 7?
A copper (II) sulphate
B aluminium sulphate
C ammonium chloride
D ethanal bisulphite
E sodium ethanoate

7 Which of the following will produce an increase in K_c for the reaction given if it occurs in an initially evacuated sealed container?

$$C(s) + H_2O(\dot{s}) \rightleftharpoons H_2(g) + CO(s)$$

A an increase in temperature
B increased surface area of carbon
C dilution of steam with nitrogen
D an increase in the partial pressure of steam
E an increase in the total pressure of the system

8 Potassium chromate can be used as an indicator when silver nitrate is titrated with sodium chloride because
A silver chromate is a coloured salt
B a colour change is occurred when $[Cl^-(aq)] = [CrO_4^{2-}(aq)]$
C potassium chromate is almost completely dissociated in aqueous sodium chloride
D K_{sp}(silver chromate) $<$ K_{sp} (silver chloride)
E $\dfrac{[Ag^+(aq)]^2_{eqm}[CrO_4^{2-}(aq)]_{eqm}}{[Ag_2CrO_4(aq)]_{eqm}} > \dfrac{[Ag^+(aq)]_{eqm}[Cl^-(aq)]_{eqm}}{[AgCl(aq)]_{eqm}}$

	STATEMENT 1	STATEMENT 2
9	The equilibrium constant for the thermal dissociation of calcium carbonate is dependent upon the partial pressure of carbon dioxide.	In heterogeneous equilibria the concentration of all pure solid phases may be considered constant.
10	The solubility product of sodium nitrate is given by $K_{sp} = [Na^+(aq)]_{eqm}\ [NO_3^-(aq)]_{eqm}$	Sodium nitrate is assumed to be totally ionized in a dilute aqueous solution.
11	Aminobenzene dissolves slightly in water to yield a solution of pH>7.	$K_b = \dfrac{[C_6H_5NH_3^+(aq)]_{eqm}\ [OH^-(aq)]_{eqm}}{[C_6H_5NH_2(l)]_{eqm}\ [H_2O(l)]_{eqm}}$
12	Aqueous solutions can only have positive valves of pH.	$pH = -\lg [H^+(aq)]_{eqm}$ and $\lg(1{\cdot}0) = 0$
13	When dinitrogen tetroxide is passed through a heated tube a colour change is observed.	Increased temperature causes dinitrogen tetroxide to dissociate and the equilibrium for the reaction is displaced to the right.
14	Magnesium hydroxide is insoluble in water but very soluble in aqueous sodium hydroxide.	Magnesium hydroxide is soluble in solutions containing a common ion such as OH^-.

15 Which of the following apply to a 'buffer solution'?

1 the pH of the solution is dependent upon the ratio of the concentrations of acid and corresponding base
2 the number of H^+ ions formed in solution from the acid in the 'buffer' is small
3 a 'buffer solution' contains a soluble salt of a weak acid
4 addition of a strong base reduces the pH of the 'buffer solution'

16 The reaction

$PCl_5(g) \rightleftharpoons PCl_3(g) + Cl_2(s)$

if taking place in an enclosed system, is one in which

1 the partial pressure of all three components is the same
2 $K_p = K_c$
3 K_p = partial pressure of any one component
4 the number of molecules present at equilibrium is greater than the number of molecules present initially

17 Which of the following apply to the equilibrium
$Br_2(l) + H_2O(l) \rightleftharpoons H^+(aq) + Br^-(aq) + HOBr(aq)$

1 $K_c = \dfrac{[HOBr(aq)]_{eqm}\ [H^+(aq)]_{eqm}\ [Br^-(aq)]_{eqm}}{[Br_2(l)]_{eqm}\ [H_2O(l)]_{eqm}}$

2 the solution is neutral because $[OH^- aq]$ from $HOBr(aq) =$ $[H^+(aq)]$ from $HBr(aq)$

3 increased $[H_2O(l)]$ increases the amount of bromine which dissolves

4 K_c has no units

18 The mass of 1 mole of sodium hydrogen sulphate can **not** be determined accurately by a freezing point (cryoscopic) determination because

1 the depression of freezing point is dependent upon the mass of solute used

2 sodium hydrogen sulphate is non-volatile

3 the vapour pressure of the solution is the same as that of the solvent

4 sodium hydrogen sulphate is almost totally ionized in dilute aqueous solution

19 A saturated aqueous solution of hydrogen chloride, in a closed system, will attain a state of equilibrium because

1 $HCl(aq) \rightleftharpoons H^+(aq) + Cl^-(aq)$

2 $H_2O(l) + H^+(aq) \rightleftharpoons H_3O^+(aq)$

3 the proton-water attraction is greater than the proton-chloride ion attraction

4 a saturated aqueous solution of hydrogen chloride is a strong acid

20 To determine K_p at equilibrium for the gaseous reaction
$CO_2(g) + NO(g) \rightleftharpoons CO(g) + NO_2(g)$
it is necessary to know the

1 value of K_c

2 temperature

3 value of the Universal Gas Constant

4 volume of the reacting gases

	STATEMENT 1	STATEMENT 2
21	The rate constant is a measure of the rate of reaction and is dependent upon the concentration of the reactants.	The greater the concentration of reactants the larger the yield of products for a reaction at equilibrium.

	STATEMENT 1	STATEMENT 2
22	A heterogeneous catalyst alters the rate of reaction and the final position of equilibrium for a given system.	The yield of sulphur trioxide from the reaction $SO_2(g) + O_2(g) \rightleftharpoons 2SO_3(g)$ is increased if vanadium (V) oxide is used as a catalyst.

23 The solubility of silver chloride is greater in a concentrated solution of sodium chloride than in water because

A a 'common ion' is present in solution

B silver chloride is a sparingly soluble electrolyte

C silver can form complex ions

D concentrated sodium chloride is strongly acidic

E ions from the silver chloride crystal lattice are attracted into solution by oppositely charged ions

24 Why is no precipitate of zinc sulphide formed when a small quantity of hydrogen sulphide is added to an acid solution of zinc sulphate?

A the dissociation of zinc sulphate is hindered in acid solution

B the sulphides of Group II metals yield hydrogen sulphide in acid solution

C the $[H^+(aq)]$ from the acid displaces the dissociation equilibrium for hydrogen sulphide to the left

D the pH of aqueous hydrogen sulphide = 7

E the solubility of zinc sulphide increases with increased pH

25 Which of the following apply to an indicator, which is a weak acid (H^+R^-), with a pH range 3·0 to 5·0?

1 $HR(aq) \rightleftharpoons H^+(aq) + R^-(aq)$

2 $pH = -\lg K_a$, when $[HR(aq)]_{eqm} = [R^-(aq)]_{eqm}$

3 any colour change which occurs depends upon the ratio

$$\frac{[HR(aq)]_{eqm}}{[R^-(aq)]_{eqm}}$$

4 the indicator is not suitable for all acid-base titrations

11 GENERAL CHEMISTRY OF METALS AND SALTS

Questions 1 to 5 refer to the list set out below. Select the response which you think is the correct answer to the question

A lithium
B calcium
C copper
D mercury
E iron

Which of the above

1 has the largest 1st ionization energy?

2 has the electronic configuration (argon core) $4s^2$ $3d^6$?

3 forms an ion in which 2 atoms of the element are combined in a covalent manner?

4 forms an oxide which reacts with excess dilute sulphuric acid to give a solution containing a sulphate of the element **and** precipitates the element from solution?

5 is used to remove water in the final stages of the preparation of pure ethanol?

6 Which of the following is the electronic configuration of a Group 2 element which forms an extremely deliquescent chloride?

A (neon core) $2s^2$
B (argon core) $4s^2$
C (argon core) $3d^{10}$ $4s^1$
D (krypton core) $4d^{10}$ $5s^2$
E (xenon core) $5d^{10}$ $4s^2$

7 All of the following complex ions can be obtained in an aqueous solution EXCEPT

A $Ca(CN)_4{}^{3-}$
B $Cu(H_2O)_4{}^{2+}$
C $Cu(NH_3)_4{}^{2+}$
D $(CH_2)N_2(CH_2COO)_4Cu^{2-}$
E $Cu\,I_4{}^{2-}$

8 Which of the following pairs of dilute aqueous solutions will precipitate an oxide when mixed at room temperature?

A mercury (II) chloride,	sodium hydroxide
B zinc nitrate,	sodium carbonate
C aluminium chloride,	sodium hydroxide
D copper (II) sulphate,	sodium carbonate
E magnesium sulphate,	sodium carbonate

	STATEMENT 1	STATEMENT 2
9	Aluminium chloride may be prepared by heating the metal in a stream of dry chlorine.	Aluminium chloride sublimes at a relatively low temperature.
10	All copper (II) complex ions are blue in colour.	A blue precipitate is observed when potassium hexacyano-ferrate (II) is added to aqueous copper (II) sulphate.
11	$Mg(s) + SiO_2(s) \xrightarrow[\text{(in inert atmosphere)}]{\Delta} Si(s) + 2MgO(s)$	Magnesium powder is a strong reducing agent.
12	When trilead tetroxide is shaken with excess dilute nitric acid lead (IV) oxide is precipitated and lead (II) nitrate remains in solution.	Trilead tetroxide reacts chemically as if it had the structure $PbO_2 . 2PbO$.
13	The oxides of all the metals which have an ultimate electron orbital containing 2 electrons are coloured.	The oxides of iron (II), manganese (II) and lead (II) are coloured.
14	When polarized light is passed through a crystal of sodium chloride the effect observed is dependent upon the orientation of the crystal.	In a sodium chloride crystal the sodium ion has a co-ordination number of 4.

15 Which of the following occur when aqueous molar solutions of ammonium sulphate and calcium nitrate are mixed at room temperature?

1 the evolution of a large amount of heat
2 an increase in pH of the solutions
3 only NH_4^+ and NO_3^- ions remain in solution
4 the solubility product of calcium sulphate is exceeded

16 $2Cu(s) + O_2(g) \rightarrow 2CuO(s)$; $-127\ kJ\ mol^{-1}$
$2C(s) + O_2(g) \rightarrow 2CO(g)$; $-137{\cdot}3\ kJ\ mol^{-1}$
$C(s) + O_2(g) \rightarrow CO_2(g)$; $-393{\cdot}4\ kJ\ mol^{-1}$
$2Ca(s) + O_2(g) \rightarrow 2CaO(s)$; $-604{\cdot}2\ kJ\ mol^{-1}$

Which of the following can be deduced from the above?

1 calcium has a greater affinity for oxygen than copper at 298 K
2 calcium will reduce copper oxide to copper on heating
3 carbon is unlikely to reduce carbon dioxide to carbon monoxide at 298 K
4 carbon dioxide is the most stable oxide, shown above, at 298 K

17 Which of the following apply to the reduction of the chromium (III) ion in an acidified aqueous solution by metallic zinc?

1 $2Cr^{3+}(aq) + Zn(s) \rightarrow 2Cr^{2+}(aq) + Zn^{2+}(aq)$
2 the reaction occurs because zinc has a greater tendency to form ions than chromium
3 the solution changes colour
4 dissolved oxygen in the water will rapidly oxidize the chromium (II) ion

18 Group 2 metals form stable halides of type $M^{2+}.2X^-$ because

1 the lattice energy of the halide crystal is large
2 the M^{2+} ion has a small radius and a high charge density
3 the first two ionization energies of the metal have low (endothermic) values
4 the combination of Group 7 elements with electrons is exothermic

19 Which of the following apply to lithium, sodium and potassium?

1 the elements form ionic hydrides
2 the chlorides are very soluble in water
3 the carbonates are stable to heat
4 the hydroxides are stable to heat

20 A metal hydroxide is precipitated and carbon dioxide is evolved when sodium carbonate is added to an aqueous solution of

1 zinc sulphate
2 lead (II) acetate
3 silver nitrate
4 iron (III) sulphate

21 Crystals of all of the following are cubic EXCEPT

A iron (III) potassium sulphate
B sodium sulphate decahydrate
C caesium chloride
D zinc blende
E sodium

22 A concentrated aqueous solution of tin (II) chloride becomes cloudy on dilution because

A $Sn^{2+}(aq) + Cl^-(aq) + H_2O(l) \rightleftharpoons Sn(OH)Cl(s) + H^+(aq)$

B $Sn^{2+}(aq) \rightleftharpoons Sn^{4+}(aq) + 2e^-$

C the degree of ionization of tin (II) chloride is dependant upon concentration

D a complex of tin (II) exists in the concentrated aqueous solution

E the $Sn(H_2O)_4^{2+}$ ion is insoluble in water

23 Pure zinc chloride may be prepared by

1 neutralizing dilute hydrochloric acid with the metal
2 heating the metal oxide in dry chlorine
3 neutralizing dilute hydrochloric acid with the metal carbonate
4 heating the metal in dry chlorine

24 Metallic silver is formed as a product when

1 silver chloride is heated with metallic zinc
2 aqueous iron (II) sulphate is added to aqueous silver nitrate
3 silver chloride is heated with sodium carbonate
4 aqueous silver ammine reacts with ethanal

	STATEMENT 1	STATEMENT 2
25	Group 2 elements are reducing agents.	Group 2 elements have only one oxidation number.

12 GENERAL CHEMISTRY OF NON-METALS

1 Sulphur combines directly with all of the following EXCEPT

A sodium
B nitrogen
C copper
D chlorine
E phosphorus

2 Which of the following is the shape of a P_4 molecule?

A (regular) tetrahedron
B cubic
C monoclinic
D (regular) hexagonal
E rhombohedral

3 The percentage of chlorine in bleaching powder may be accurately determined by

A titrating bleaching powder with potassium permanganate
B titrating bleaching powder, directly, with silver nitrate
C adding excess of iron (II) sulphate to bleaching powder and titrating with potassium dichromate
D adding excess of acidified potassium iodide to bleaching powder and titrating with sodium thiosulphate
E titrating bleaching powder, directly, with sodium hydroxide

4 Which of the following yield oxygen on strong heating?

A hydroxylamine
B lithium hydroxide
C trilead tetroxide
D potassium hypochlorite
E sodium thiosulphate

5 Which of the following is a correct representation of a CO_3^{2-} ion?

A 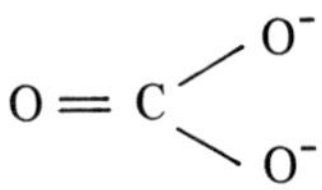

B 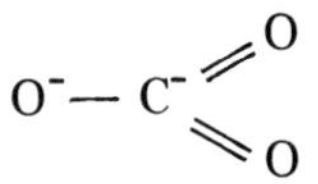

C 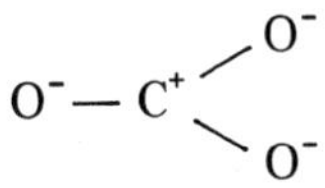

D 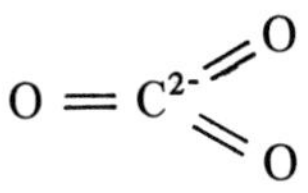

E 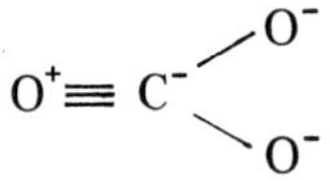

6 Dinitrogen oxide is liberated as an *initial* gaseous product when

A a strongly acid solution of a nitrate is warmed with aqueous iron (II) sulphate

B bromine is added to an acid solution of a nitrite

C zinc dust is added to moderately concentrated nitric acid

D hydrogen sulphide is passed into concentrated nitric acid

E aqueous sodium sulphite is heated with moderately concentrated nitric acid

7 Ammonia acts as a reductant in all of the following EXCEPT

A $3PbO(s) + 2NH_3(g) \rightarrow 3Pb(s) + N_2(g) + 3H_2O(l)$

B $NH_3(g) + 3Cl_2(g) \rightarrow NCl_3(l) + 3HCl(g)$

C $2Na(s) + 2NH_3(g) \rightarrow 2NaNH_2(s) + H_2(g)$

D $4NH_3(g) + 3O_2(g) \rightarrow 2N_2(g) + 6H_2O(l)$

E $8NH_3(g) + 3Cl_2(g) \rightarrow 6NH_4Cl(s) + N_2(g)$

8 In which of the following reactions is hydrogen peroxide the oxidant?

A $H_2O_2(aq) + 2H^+(aq) + 2I^-(aq) \rightarrow 2H_2O(l) + I_2(s)$

B $H_2O_2(aq) + Ag_2O(s) \rightarrow 2Ag(s) + H_2O(l) + O_2(g)$

C $PbO_2(s) + H_2O_2(aq) + 2OH^-(aq) \rightarrow Pb^{2+}(aq) + 2H_2O(l) + O_2(g)$

D $2Fe^{3+}(aq) + H_2O_2(aq) + 2OH^-(aq)$
$\rightarrow 2Fe^{2+}(aq) + 2H_2O(l) + O_2(g)$

E $5H_2O_2(aq) + 2MnO_4^-(aq) + 6H^+(aq)$
$\rightarrow 2Mn^{2+}(aq) + 8H_2O(l) + 5O_2(g)$

	STATEMENT 1	STATEMENT 2
9	$SO_3^{2-}(aq) + 5H_2O_2(aq) + I_2(s) \rightarrow SO_4^{2-}(aq) + 2I^-(aq) + 2H^+(aq)$	Iodine is more electronegative than sulphur.
10	Sulphur forms long chain molecular crystals.	Sulphur melts at a relatively low temperature to form a viscous liquid which changes its physical characteristics with change in temperature.
11	Platinum electrodes are suitable for the electrolysis of acidified water.	For the electrode system represented by $2H^+(aq) \mid [H_2(g)]$ Pt; $E^{\ominus}_{298K} = 0{\cdot}00$ V.
12	Ammonia is displaced from solution when hot equimolar aqueous solutions of disodium hydrogen phosphate and ammonium chloride are mixed.	An 0·1M aqueous solution of orthophosphoric acid is a stronger acid than 0·1M aqueous hydrogen chloride.

	STATEMENT 1	STATEMENT 2
13	More energy is lost to the surroundings when 1 mole of hydrogen sulphide is formed from its elements than when 1 mole of water is formed.	Hydrogen sulphide is oxidized more easily than water.
14	Fluorine may be prepared by the action of heat on a mixture of sodium fluoride, manganese (IV) oxide and concentrated sulphuric acid.	When concentrated sulphuric acid is heated with a sodium halide a volatile, covalent hydrogen halide is displaced from the mixture.

15 The solution obtained, when aqueous sodium hydroxide is added dropwise to bromine water until the colour is just discharged, will

1 precipitate manganese (IV) oxide from aqueous manganese (II) sulphate
2 precipitate lead (IV) oxide from aqueous lead (II) nitrate
3 oxidize aqueous sulphur (IV) ions to sulphur (VI) ions
4 not give a precipitate with aqueous acidified silver nitrate

16 Boron is a non-metal which

1 forms an ammine, $BCl_3 . NH_3$
2 does not combine directly with nitrogen
3 forms an acid, H_3BO_3
4 does not decompose steam, even at red heat

17 Elements of atomic number 15, 16 and 17 will

1 form oxoacids of type, HXO_3
2 form covalent compounds with oxygen
3 not displace aluminium from its salts
4 attain the krypton configuration when combined with other elements

18 Which of the following statements are correct?

1 $2F_2(g) + 2OH^-(aq) \rightarrow 2F^-(aq) + F_2O(g) + H_2O(l)$
2 fluorine is reduced by the addition of oxygen
3 sodium fluoride is non-poisonous
4 fluorine does not combine directly with lead

19 In which of the following does the oxidation number of sulphur correctly correspond to the given formula?

	Formula	*Oxidation number*
1	H_2SO_3	+4
2	$H_2S_2O_3$	+2
3	S_8	0
4	H_2S	-2

20 Which of the following apply to *all* non-metals?

1 do not give a conducting solution even if soluble in water
2 will not dissolve in organic solvents
3 exist in more than one state at the same temperature
4 can form covalent bonds with other elements

	STATEMENT 1	STATEMENT 2
21	$Si(s) + H_2O(aq) + 2OH^-(aq) \rightarrow SiO_3^{2-}(aq) + 2H_2(g)$	Silicon is in the same period as aluminium.
22	Anhydrous nitric acid may be obtained by fractional distillation of an aqueous solution of nitric acid.	Nitric acid and water will not form a constant boiling point (azeotropic) mixture.

23 Sodium sulphite will

1 liberate oxygen when strongly heated
2 be oxidized to the sulphate when oxygen is passed through a hot aqueous solution
3 be reduced to sulphur on additon of a concentrated mineral acid
4 yield sodium thiosulphate when boiled with sulphur

24 Which of the following are necessary when hydrogen peroxide is prepared from barium peroxide and syrupy orthophosphoric acid?

1 the orthophosphoric acid should be diluted
2 barium peroxide should be added slowly to the acid
3 the temperature should be maintained as low as possible
4 solid residues should be discarded after filtration

25 A co-ionic bond is present in

A phosphorus (V) oxide
B phosphine
C phosphonium iodide
D trimethyl phosphine
E phosphorus (III) chloride

13 ALKANES, ALKENES AND ALKYNES

Questions 1 to 5 refer to the list of formulae given below. Select the response which you think answers the question correctly.

A 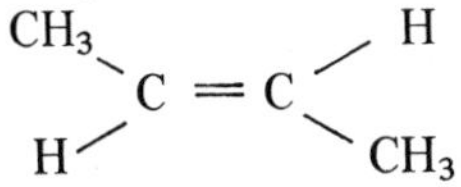

B 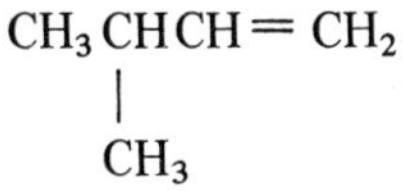

C 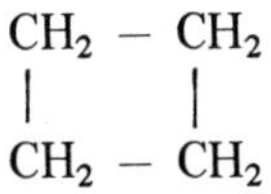

D 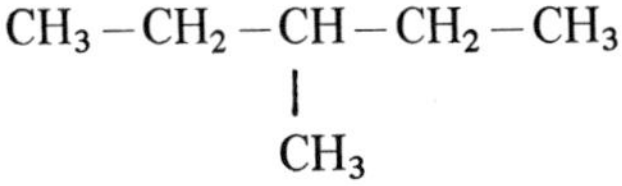

E $CH_3CH_2C \equiv CH$

Which of the above

1 is a planar, highly strained, molecule?

2 is trans-but-2-ene?

3 contains one σ and two π bonds between two adjacent carbon atoms?

4 contains four secondary (2°) groups?

5 is a compound in which groups at the extreme ends of the molecule are capable of rotating freely with respect to each other?

6 The formula of the compound which may be prepared by the electrolysis of an aqueous solution of sodium propanoate is

A CH_3CH_3
B $CH_2 = CH_2$
C $CH_2 = CH\,CH_2CH_3$
D $CH_3CH = CH\,CH_3$
E $CH_3CH_2CH_2CH_3$

7 Which of the following products, obtained by the fractional distillation of petroleum, contains the largest percentage of carbon?

A petroleum ether
B light naphtha
C gasoline
D kerosene
E gas oil

8 Which of the following compounds is formed when sodium reacts with an ethereal solution of an alkyl halide?

A n-propane
B n-butane
C 2-methylhexane
D ethene
E but-2-yne

9 Alkanes may be synthesized by

1 hydrogenation of an alkyl halide using a zinc/copper couple in aqueous alcohol
2 acid hydrolysis of an alkyl magnesium halide
3 passing alkene vapour and hydrogen over a hot platinum catalyst
4 reduction of an acid by an ethereal solution of lithium aluminium hydride

10 What happens when equimolecular volumes of chlorine and ethyne are mixed in the presence of a metal halide catalyst?

1 the gases react explosively
2 an addition reaction occurs
3 hydrogen chloride is evolved
4 1, 2-dichloroethane is produced

11 Which of the following methods are suitable for the production of ethyne?

1 the pyrolysis of methane
2 acid hydrolysis of calcium carbide
3 heating 1, 2-dibromoethane with ethanolic potassium hydroxide
4 warming 1, 2-dibromoethane vapour with soda lime

12 Alkenes undergo an addition reaction with

1 cold concentrated sulphuric acid
2 potassium permanganate in acid solution
3 ozone
4 hydrogen halides

13 When ethene is shaken with dilute alkaline potassium permanganate then

1 ethane-1, 2-diol is produced
2 the solution is decolourized
3 the MnO_4^- ion is reduced
4 the oxidation number of manganese changes by +5

14 The gases produced when propane is pyrolysed are

1 propene
2 ethyne
3 hexane
4 hydrogen

	STATEMENT 1	STATEMENT 2
15	Alkanes undergo chlorination in the presence of ultra-violet light.	Chlorine molecules undergo homolytic fission when irradiated by ultra-violet light.
16	Wet liquid bromine undergoes electrophilic addition with alkenes.	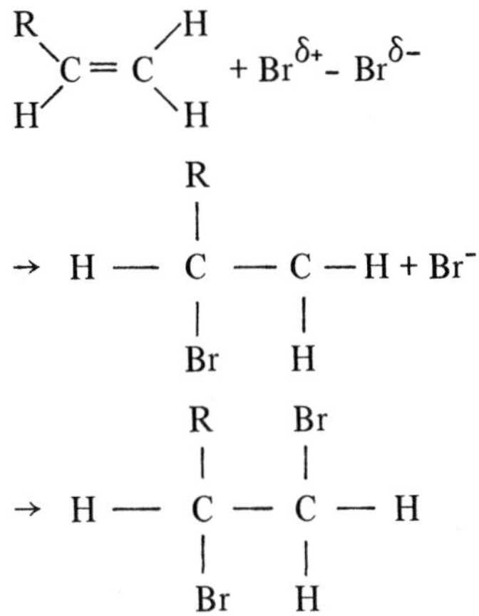
17	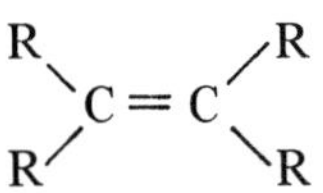 is less stable to electrophilic addition than 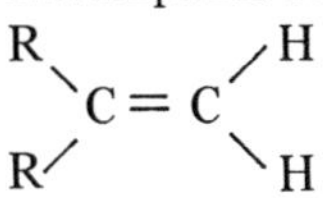	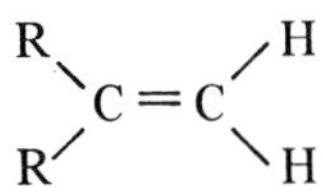 is an isomer of 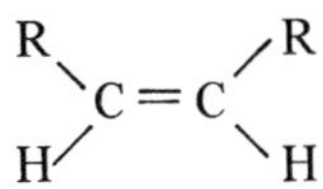
18	The boiling point of butane is lower than that of octane.	Butane contains three 2° carbon atoms.
19	In an alkene at least one carbon atom exhibits sp hybridization.	Carbon and hydrogen atoms in ethyne are linked by π bonds.
20	But-2-ene undergoes homolytic fission on irradiation by ultra-violet light.	C=C bonds are weaker than C–C bonds.

21 But-1-yne may be prepared by

A the alcoholysis of 1,2-dibromobutane
B the dehydration of butane-1,4-diol
C catalytic pyrolysis of butane
D polymerization of ethene in the presence of a silver catalyst
E the hydrolysis of 1,4-dibromobutane

22 Which of the following methods can be used for the preparation of methane?

A the decarboxylation of sodium ethanoate
B the electrolysis of aqueous sodium ethanoate
C the hydrolysis of cyclohexane
D the hydrolysis of an ozonide
E the alcoholysis of ethene oxide

23 A higher alkane is formed when

A alkenes are hydrogenated in the presence of a silver catalyst
B an alkane is added to an alkene in the presence of an acid catalyst
C an alkane is refluxed with sulphuryl chloride in the presence of a catalyst
D ethyne is passed into dilute sulphuric acid containing mercury (II) sulphate
E acetylene dimerizes in an acid solution of copper (I) chloride and ammonium chloride

24 Which of the following compounds has a puckered structure known as a *chair conformation*?

A 2-methylbutane
B 2-methylbut-1-ene
C cyclohexane
D 2-methyl-4-ethylpentane
E *trans*-1-methylpropene

	STATEMENT 1	STATEMENT 2
25	The red colour of bromine water is discharged on the addition of an alkene.	Bromine forms a σ bond with an unsaturated carbon atom and the Br-OH bond breaks.

14 ALCOHOLS AND ALKYL HALIDES

1 Tri-iodomethane will be produced when an alkaline aqueous solution of iodine is warmed with

A propan-1-ol
B butan-1-ol
C propan-2-ol
D 2-methyl-propan-2-ol
E methanol

2 A principal product formed when the vapour of propan-2-ol is passed over hot copper turnings is

A propane
B propene
C methanal
D ethanol
E propanone

3 The -OH group in a primary alcohol is replaced by chlorine in all of the following reactions EXCEPT

A passing chlorine directly into the hot alcohol
B passing dry hydrogen chloride into the alcohol in the presence of anhydrous zinc chloride
C heating the alcohol with phosphorus (V) chloride
D heating the alcohol with phosphorus (III) chloride
E heating the alcohol with thionyl chloride in the presence of pyridine

4 In which of the following is an ether a principal product?

A the alkaline alcoholysis of ethanol
B refluxing ethanol and concentrated sulphuric acid at a temperature above 450 K
C the hydrolysis of bromoethane by dry silver oxide
D refluxing iodoethane with ethanolic silver nitrite
E the addition of red hot copper (II) oxide to propan-2-ol

5 Which of the following will *most* rapidly undergo nucleophilic substitution on hydrolysis with pure water?

A CH_3CH_2Br
B $CH_3CHBrCH_3$
C $CH_3C(CH_3)_2Br$
D CH_3CH_2ONa
E $CH_2NO_3CHNO_3CH_2NO_3$

6 Which of the following methods is *most* suitable for the preparation of the compound of formula

$$CH_3COOCH(CH_3)_2$$

(structure printed as: CH_3 above, CH_3COOCH, CH_3 below, joined by vertical bonds to the CH)

A acylation of 2-chloropropane
B refluxing an ethanolic solution of 2-chloropropane and sodium ethoxide
C refluxing 2-chloropropane with ethanolic ammonia
D refluxing 1, 2-dichloropropane with ethanolic potassium cyanide
E refluxing an ethanolic solution of 2-chloropropane and silver ethanoate

7 When ethane-1, 2-diol is heated with potassium hydrogen sulphate a vapour is evolved which will give the haloform reaction with alkaline aqueous iodine. The vapour is

A ethanol
B ethanal
C propanone
D methanal
E acrolein

8 Which of the following are characteristic of tri-chloromethane?
A hydrolysed by ethanolic potassium hydroxide to give potassium methanoate
B precipitation on warming with aqueous silver nitrate
C addition with an ethereal suspension of magnesium
D nucleophilic substitution with ethanolic sodiumethoxide
E alkylation with ammonia to yield a 3° amine

	STATEMENT 1	STATEMENT 2
9	$CH_3CH_2Br + AgCN(alc) \rightarrow CH_3CH_2NC + AgBr$	Silver is attached to the nitrogen atom in silver cyanide.
10	The rate of hydrolysis of 1° alkyl halides is dependent only upon the concentration of the alkyl halide.	Hydrolysis of 1° alkyl halides is a nucleophilic, unimolecular substitution reaction.

	STATEMENT 1	STATEMENT 2
11	The compound of formula 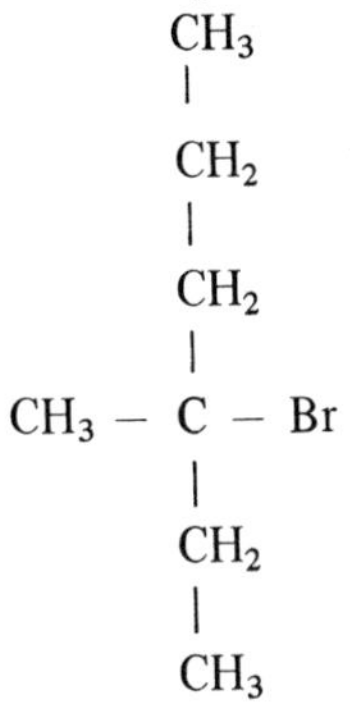 exhibits optical isomerism	An optically inactive mixture is obtained when 3-bromo-3-methyl hexane is hydrolysed.
12	Butan-2-ol will yield a mixture of two isomeric alkenes when heated with concentrated sulphuric acid above 450 K.	$CH_3CH_2CH(OH)CH_3 + H^{\delta+}-OS^{\delta-}O_2\,OH \rightleftharpoons CH_3CH_2CH(OH_2^+)CH_3 + HSO_4^-$
13	Alkoxides are crystalline solids readily soluble in water to yield an alkaline solution.	Alcohols are weaker acids than water.
14	A convenient way to distinguish between 1°, 2° and 3° alcohols is to oxidize the alcohol in alkaline media and then test for the presence of an aldehyde.	1° alcohols yield aldehydes on oxidation, 2° alcohols give ketones, but 3° alcohols are stable to oxidation by alkaline permanganate.

15 Which of the following occur when bromoethane is prepared from red phosphorus, potassium bromide, ethanol and concentrated sulphuric acid?

1 $2KBr(s) + H_2SO_4(l) \rightarrow K_2SO_4(s) + 2HBr(g)$
2 protonation of the alcohol
3 elimination of a molecule of water from the $CH_3CH_2OH_2^+$ ion
4 an addition reaction between a nucleophile and an electrophile

16 Absorption peaks corresponding to which of the following, obtained on an infra red absorption spectrum for ethanol, may be considered characteristic of a primary alcohol?

1 C–H (stretching)
2 C–H (bonding)
3 O–H (stretching)
4 C–OH (stretching)

17 Peaks corresponding to the presence of which of the following ions may be shown on a mass spectrum of trichloromethane?

1 C^+
2 CH^+
3 $CHCl^+$
4 $CHCl_2^+$

18 Grignard reagents can be used to prepare

1 1° alcohols from methanol
2 2° alcohols from propanone
3 2° alcohols from ethanol
4 3° alcohols from ethoxyethane

19 Which of the following apply to the industrial manufacture of methanol?

1 $2H_2(g) + CO(g) \rightleftharpoons CH_3OH(g)$
2 the forward reaction proceeds more efficiently at a high temperature
3 the yield is increased by application of a high pressure
4 the reaction is catalysed by the oxides of zinc and chromium (III)

20 Which of the reductions, given below, can be brought about by lithium aluminium hydride?

1 $C_2H_5COOC_2H_5 \rightarrow 2C_2H_5OH$
2 $C_2H_5COOH \rightarrow C_2H_5CH_2OH$
3 $\begin{matrix} C_2H_5 \\ H \end{matrix}\!\!>C{=}O \rightarrow C_2H_5OH + CH_3OH$
4 $\begin{matrix} C_2H_5 \\ C_2H_5 \end{matrix}\!\!>C{=}O \rightarrow \begin{matrix} C_2H_5 \\ C_2H_5 \end{matrix}\!\!>CHOH$

21 The alcohol obtained on an industrial scale by direct hydrolysis of an alkene is

A methanol
B ethanol
C propan-2-ol
D butan-2-ol
E 2,2-dimethylethanol

22 The formula of the compound which is readily oxidized to carbon dioxide and water on warming with acidified dilute potassium permanganate is

A $(CH_3)_3CBr$ **D** CHI_3

B $(CH_3)_3COH$ **E** $HOCH_2CH_2OH$

C CH_3CHCl_2

23 2-methylpropan-1-ol can be synthesized from

A 2-methylbromopropane and aqueous silver cyanide

B 2-methylbromopropane and ammoniacal silver nitrate

C 2-methylpropanone and ethyl magnesium bromide

D 2-methylpropoxide and water

E ethyl butanoate and aqueous sodium hydroxide

	STATEMENT 1	STATEMENT 2
24	Propane-1,2,3-triol is a secondary alcohol.	Propane-1,2,3-triol yields 1,2,3-trichloropropane on reaction with phosphorus (V) chloride.
25	1° alkyl iodides are less reactive than alkyl fluorides.	1° alkyl halides rapidly undergo nucleophilic substitution.

15 ARENES AND ARYL COMPOUNDS

1 Direct substitution into the benzene ring may be carried out by

A passing chlorine into boiling toluene
B heating phenol and hydrogen chloride
C refluxing toluene and hydrogen chloride
D passing chlorine into hot benzene in the presence of a halogen carrier
E passing chlorine into benzene irradiated by ultra-violet light

2 Which of the following occur when phenol is added to aqueous bromine?

A a red precipitate is formed
B bromine is substituted for the hydroxyl group
C 2,4,6-tribromophenol is formed
D 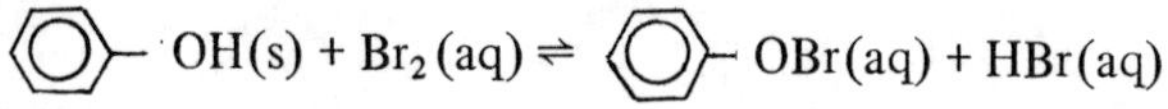

E a mixture of ortho- and para-bromophenols are formed

3 Which of the following apply to the laboratory preparation of nitrobenzene from a mixture of concentrated nitric and sulphuric acids and benzene?

A nitric acid acts as a base which reacts with the sulphuric acid to yield the NO_2^+ ion
B concentrated sulphuric acid is a dehydrating agent
C the rate of reaction is dependent upon the concentration of benzene
D sulphuric acid inhibits the formation of ions leaving free dinitrogen tetroxide available
E the addition of sulphuric acid to nitric acid is exothermic and enables the reaction to proceed

4 Phenol is a principal product in each of the following reactions EXCEPT

A the acid hydrolysis of sodium phenoxide
B the action of heat on an aqueous benzenediazonium salt
C the fusion of chlorobenzene and sodium hydroxide
D the fusion of sodiumbenzene sulphonate and sodium hydroxide
E the catalytic hydrolysis of chlorobenzene at 700 K

5 What happens when a small quantity of benzoic acid crystals are gently warmed in a test-tube?

A charring, leaving a carbon residue

B decarboxylation, leaving benzene

C sublimation, crystals forming in the cooler region of the test-tube

D condensation, producing an acid anhydride

E thermal decomposition, to yield a mixture of lower hydrocarbons and carbon dioxide

6 The chlorine atom present in chlorobenzene is ortho- and para-directing with respect to further substitution because

A chlorine withdraws electrons from the ring

B chlorine activates the ring with respect to electrophilic substitution

C nitrating agents produce a mixture of ortho- and para-derivatives

D the carbon atom combined with the chlorine is sp hybridized

E chlorine is an electronegative element

7 Which of the following occur when concentrated hydrochloric acid is added dropwise to a suspension of aniline in water?

A a white precipitate is formed

B a vigorous exothermic reaction

C $C_6H_5NH_2(l) + HCl(aq) \rightleftharpoons C_6H_5NH_3^+(aq) + Cl^-(aq)$

D increased H^+ concentration promotes the reaction $C_6H_5NH_2(l) + H_2O(aq) \rightarrow C_6H_5NH_3^+(aq) + OH^-(aq)$

E the dissociation constant for water is exceeded and phenylamine then dissolves

8 Which of the compounds listed below forms a dimer with two hydrogen bonds between each molecule, in the vapour phase?

A acetophenone

B phenol

C acetanilide

D methyl phenyl ether

E benzoic acid

	STATEMENT 1	STATEMENT 2
9	The -CHO group deactivates the ring and is meta-directing with respect to further substitution.	The carbon atom in a carbonyl group is electron deficient because the carbonyl group has a strong polar character.

	STATEMENT 1	STATEMENT 2
10	Methyl phenyl ether is more reactive than ethyl methyl ether.	Methyl phenyl ether yields phenol when refluxed with excess hydrogen bromide.
11	Benzal is oxidized to potassium benzoate on warming with saturated aqueous potassium hydroxide.	Aromatic aldehydes are more readily oxidized than aliphatic aldehydes.
12	A diazonium ion in aqueous solution readily undergoes nucleophilic attack on warming.	$C_6H_5N_2^+(aq) \xrightarrow{\Delta} C_6H_5^+(aq) + N_2(g)$
13	Electrophilic substitution of phenol takes place more readily in acid than in alkaline solution.	Phenol forms phenoxide ions in alkaline solution.
14	The benzene ring is activated when electron donating groups are substituted for a hydrogen atom.	Ortho- and para-derivatives are obtained when an activated ring reacts with electrophilic groups.

15 The reagents required for the laboratory preparation of free aminobenzene are

1 nitrobenzene
2 tin
3 concentrated hydrochloric acid
4 sodium hydroxide pellets

16 Which of the following occur in the benzene nucleus, according to the molecular orbital theory (or modern bonding theory)?

1 sp^2 hydridization of the carbon atoms
2 carbon atoms are σ-bonded to each other
3 p-orbitals are situated perpendicular to the plane of the molecule
4 the angle between a C–C and an adjacent C–H bond is 120°

17 Which of the following demonstrate the acidic character of benzoic acid?

1 hydrogen is liberated when the aqueous acid is warmed with magnesium
2 hydrogen chloride is evolved on addition of phosphorus halides
3 carbon dioxide is displaced from carbonates by the aqueous acid
4 the acid is very soluble in water

18 Which of the following reactions of phenylmethylamine are characteristic of alkyl amines?
1 hydrolysis with water
2 acylation with an ethanoyl chloride
3 esterification with ethanolic silver nitrate
4 salt formation with dilute aqueous sulphuric acid

19 Which of the following groups are meta-directing?

1 -NHR
2 $-SO_3H$
3 -OR
4 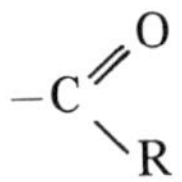

20 Benzal will
1 slowly turn Schiff's reagent purple
2 reduce Fehlings solution
3 decolourize acidified potassium permanganate
4 form a resin with a saturated solution of an aqueous alkali

Questions 21 to 25 refer to the list given below. Select the response which you think answers the question correctly.

A alkylation
B hydrogenation
C saponification
D hydrolysis
E acylation

Which of the above 'reaction types' can be used to describe the method of preparation of

21 acetophenone, from benzene and one other compound only, using anhydrous aluminium chloride as a catalyst?

22 methylcyclohexane, from methylbenzene and one other substance only, using nickel as a catalyst?

23 methylbenzene, from benzene and one other substance only, using anhydrous aluminium chloride as a catalyst?

24 benzene from benzene sulphonic acid by warming with dilute aqueous sulphuric acid?

25 benzoic acid by warming an alkaline aqueous solution of benzamide?

16 COMPOUNDS CONTAINING THE FUNCTIONAL GROUPS

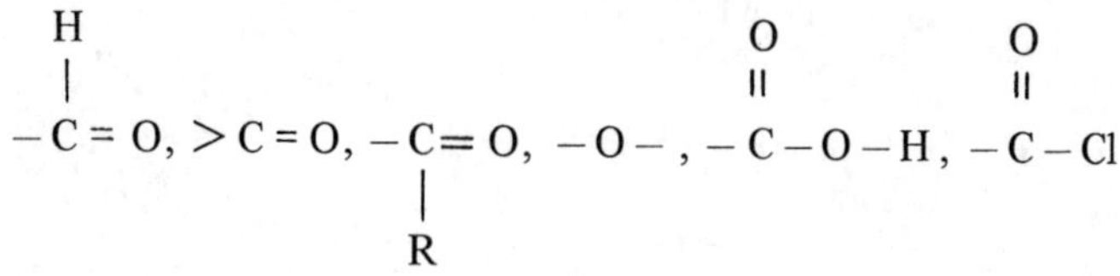

Questions 1 to 5 refer to the formulae listed below. Select the response which you think answers the question correctly.

A CH_3CHO **D** CH_3COOH
B $(CH_3)_2CO$ **E** CH_3COCl
C $(CH_3)_2O$

Which of the above is the formula of the compound which
1 undergoes exothermic polymerization on the addition of a few drops of concentrated sulphuric acid?

2 will bring about the reduction $Ag^+(aq) \rightarrow Ag(s) + e$?

3 is decomposed by the action of cold water?

4 may be prepared by bubbling carbon dioxide into an ethereal solution of a Grignard reagent and then hydrolysing the product?

5 forms an unstable peroxide when irradiated by ultra-violet light?

6 Methanoic acid is a stronger acid than ethanoic acid because
A there is a mesomeric shift of electrons as shown

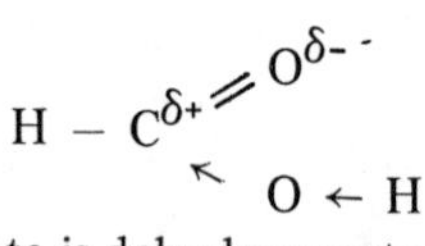

B sodium methanoate is dehydrogenated on heating to form sodium ethanedioate
C the inductive effect of the CH_3- group is greater than the inductive effect of a hydrogen atom
D the carboxylate ion forms two resonance hybrids
E an equimolar solution of methanoic acid and dilute aqueous ammonia will give a precipitate with neutral iron (II) chloride

7 Which of the following apply to the formation of white crytsals when excess concentrated aqueous ammonia has been added to ethanal?

A 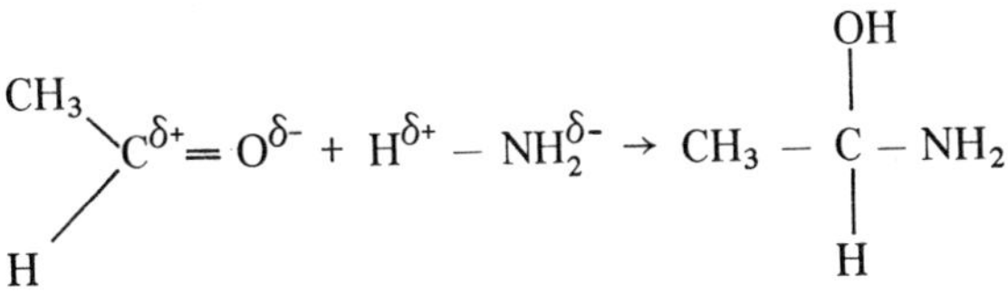

B a complex condensation reaction has occurred

C ethanol polymerizes

D $NH_3(aq) + H_2O(aq) \rightleftharpoons NH_4^+(aq) + OH^-(aq)$

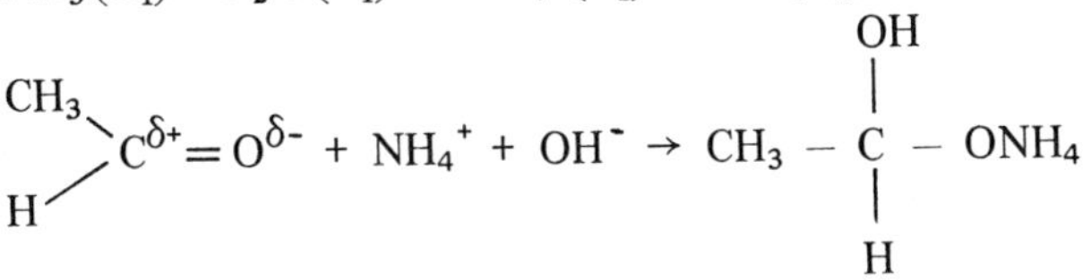

E a resin is formed which has a characteristic odour

8 Methylphenylketone will react in a different manner than propanone with

A ammoniacal silver nitrate

B acidified potassium permanganate

C Schiff's reagent

D Fehling's solution

E aqueous alkaline iodine

	STATEMENT 1	STATEMENT 2
9	Hydrolysis of an ester is best brought about under alkaline conditions.	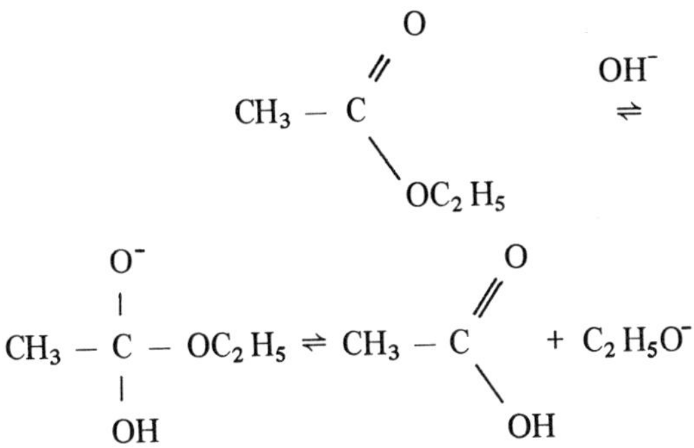
10	Propanone will form an ester when refluxed with propanoic acid.	Propanone contains an acyl group and will form esters in the same way as ethanoyl chloride.

	STATEMENT 1	STATEMENT 2
11	Methanoic acid is a reducing agent.	A white precipitate forms when methanoic acid is added to warm aqueous mercury (II) chloride.
12	Acid chlorides are more susceptible to nucleophilic attack than acid anhydrides.	Ethanoic anhydride will not react with phenylamine to yield phenylmethanamide.
13	The double bond, C = O, in propanone acts as a source of electrons and is susceptible principally to electrophilic attack.	$(CH_3)_2C{=}O + PCl_5 \rightarrow CH_3-CCl_2-CH_3 + POCl_3$
14	Trichloroethanal reacts with water to form a crystalline hydrate.	The chlorine atoms in trichloroethanal withdraw electrons from the carbonyl group making the carbonyl carbon more positive than in propanone.

15 Which of the following is a possible resonance structure for benzoyl chloride?

1

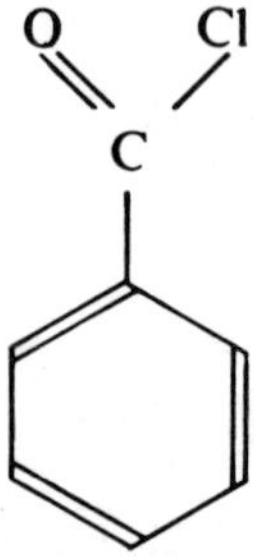

2

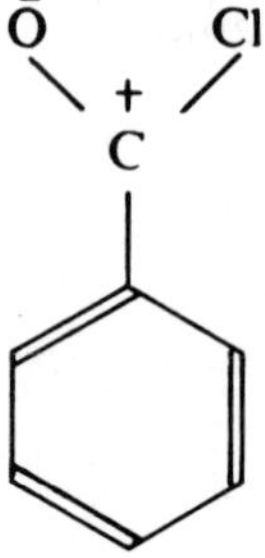

3

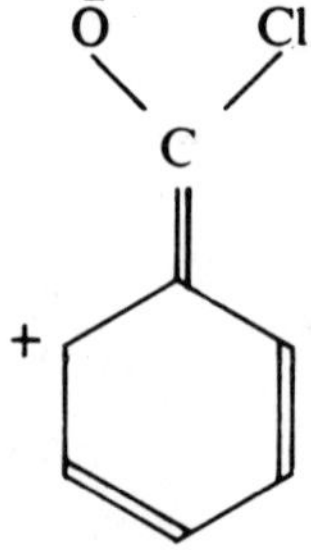

4

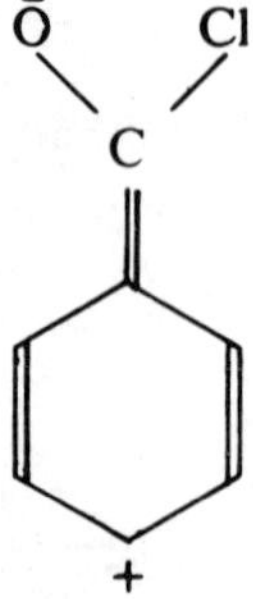

16 Which of the following is used to test for the presence of a reactive carbonyl group in an organic compound?

1 sodium hydrogen sulphite
2 hydroxylamine hydrochloride
3 lithium aluminium hydride
4 2,4-dinitrophenylhydrazine

17 Acid chlorides will undergo a condensation reaction with

1 an alcohol
2 an amine
3 ammonia
4 benzene

18 Aldehydes *and* ketones may be prepared by

1 oxidation of an alcohol
2 reduction of an acid chloride
3 decarboxylation of the sodium salt of an organic acid
4 addition-elimination reaction between a Grignard reagent and an alkyl cyanide

19 Which of the following apply to ethanoic acid?

1 the vapour decomposes when the heated vapour passes over a nickel catalyst
2 chlorination occurs when chlorine is bubbled into the hot acid in the presence of red phosphorus
3 forms a sodium salt and an alcohol when refluxed with saturated aqueous sodium hydroxide
4 the acid undergoes nucleophilic substitution when heated with phosphorus (V) chloride

20 An ether may be prepared by

1 refluxing a potassium alkoxide with an alkyl halide
2 treating a higher alkyl amine with warm acidified sodium nitrite
3 warming an alkyl halide with phenol in alkaline solution
4 warming an alkyl halide with ethereal silver nitrite

21 Ethanal may be converted to ethanoic anhydride by

A catalytic oxidation, by cobalt and copper acetate catalysts
B light catalysed autoxidation
C intermolecular condensation, caused by strong heating
D refluxing with dilute aqueous alkali
E refluxing with concentrated sulphuric acid

22 Benzoic acid will react with all of the following EXCEPT

A magnesium
B phosphorus (V) chloride
C iron (III) chloride
D benzoyl chloride
E tin (II) chloride

23 The most probable mechanism for the reaction between a ketone and a primary amine is

A 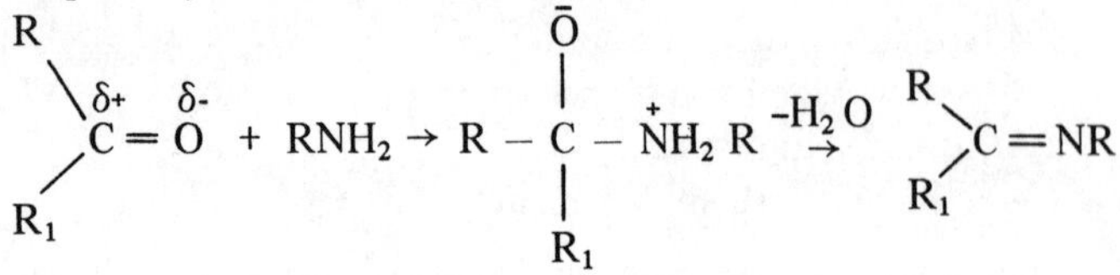

B 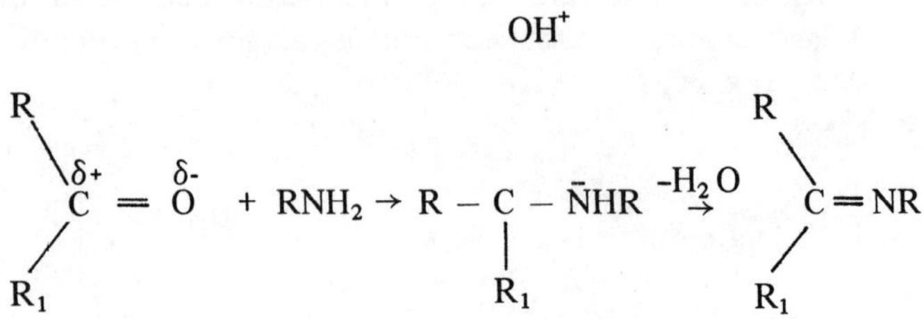

C 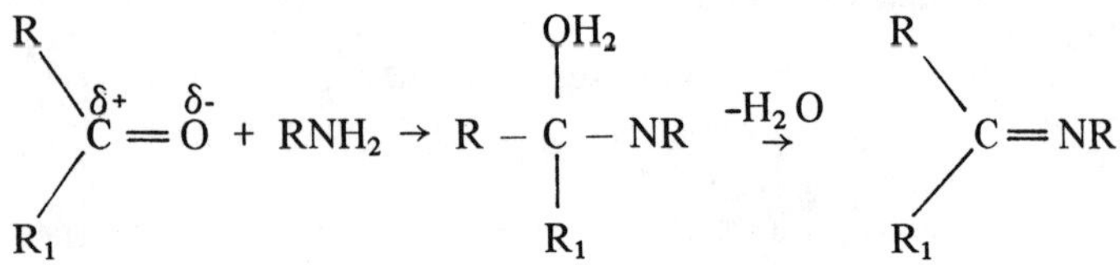

D 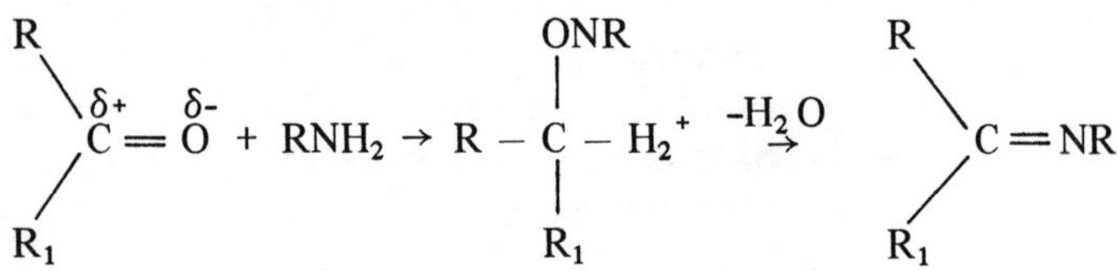

E 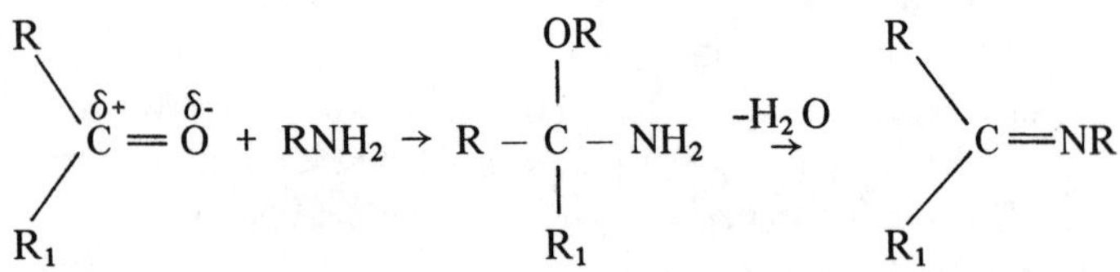

24 Which of the following apply to aliphatic ethers?

1 form a homologous series

2 have the molecular formula $C_n H_{2n+2} O$

3 are isomeric with aliphatic monohydric alcohols

4 do not exist as cyclic compounds

	STATEMENT 1	STATEMENT 2
25	Aliphatic ketones will not reduce Fehling's solution.	Propanone is less electrophilic than ethanal.

17 COMPOUNDS CONTAINING THE GROUPS $-CH$, $-NC$, $-NH_2$, $-NHR$, $-NR_2$, $-NO_2$, $-CONH_2$

1 Which of the following is the formula of the compound formed when bromine reacts with an aqueous alkaline solution of an acid amide?

A CBr_3CONH_2
B CH_3CBr_2
C CH_3NH_2
D $CH_3CONHBr$
E CH_3CN

2 The formula of the compound formed when aqueous potassium cyanide is refluxed with iodopropane is

A CH_3CH_2CN
B $CH_3CH_2CO_2H$
C $CH_3CH_2NH_2$
D $CH_3CH_2COONH_4$
E CH_3CH_2OH

3 The products formed when urea reacts with acidified sodium nitrite are

A carbon dioxide, nitrogen and water
B glycine and ammonia
C methanal, nitrogen and water
D methanoic acid, ammonia and water
E a compound containing a peptide linkage and ammonia

4 Alkyl nitrites will yield

A a primary amine with sodium and ethanol
B an alcohol with a zinc/copper couple in aqueous ethanol
C a carboxylic acid on heating with acidified potassium permanganate
D a diazo compound with cold acidified sodium nitrite
E ammonia on heating with aqueous sodium hydroxide

5 Which of the following is the structural formula of nitroethane?

A $CH_3-CH_2-N\begin{smallmatrix} \nearrow\!\!\!\!/\, O \\ \searrow O \end{smallmatrix}$ (N doubly bonded to one O, singly bonded to the other O)
B $CH_3-CH_2-O-N=O$
C $CH_3-CH_2\ N=O$
D $CH_3-CH_2-N \rightarrow O$
E $CH_3-CH_2-N\begin{smallmatrix} \diagup O^{\delta+} \\ \diagdown O^{\delta-} \end{smallmatrix}$

6 A compound of molecular formula C_3H_5N is formed when

A nitropropane is reduced with tin and hydrochloric acid
B ammonium ethanoate is refluxed with phosphorus (V) oxide
C propanal is refluxed with aqueous potassium cyanide
D propanal vapour, hydrogen and ammonia are passed over a heated nickel catalyst
E a mixture of ethylamine, trichloromethane and ethanolic potassium hydroxide are warmed

7 Which of the following forms a soluble salt on treatment with nitrous acid?

A trimethylamine
B ethanamide
C propionitrite
D nitromethane
E methylammoniumchloride

8 Ethanamide will undergo all the reactions represented by the following equations, under normal laboratory conditions EXCEPT

A $2CH_3CONH_2(s) + 2Na(s) \xrightarrow{\Delta} 2CH_3CONHNa(s) + H_2(g); -\Delta H$
B $CH_3CONH_2(aq) + H^+(aq) \rightarrow CH_3CONH_3^+(aq)$
C $CH_3CONH_2(s) + C_2H_5OH(l) \rightleftharpoons CH_3COOC_2H_5(l) + NH_3(g)$
D $2CH_3CONH_2(aq) + HgO(s) \rightarrow (CH_3CONH)_2Hg(aq) + H_2O(l)$
E $CH_3CONH_2(s) + CH_3COONH_4(s)$
$\rightarrow (CH_3CO)_2O(l) + 2NH_3(g); +\Delta H$

	STATEMENT 1	STATEMENT 2
9	All aliphatic amines have a characteristic 'fishy' odour and are gaseous at room temperature.	The product formed when iodomethane is heated in a sealed tube with ammonia is gaseous and has a characteristic odour.
10	The 2p electrons of the nitrogen atom form covalent bonds with alkyl radicals in a 3° amine.	The nitrogen atom is sp^3 hybridized and the 2s electrons form the lone pair of electrons on the nitrogen atom.
11	$(CH_3)_3N + CH_3I \xrightarrow{\Delta} (CH_3)_4N^+I^-$	Heat is required for the reaction between a tertiary amine and an alkyl halide to proceed as tertiary amines are not nucleophiles.

	STATEMENT 1	STATEMENT 2
12	Tetraethylammonium iodide can be prepared by heating iodoethane with alcoholic ammonia.	Ethylamine is more reactive than ammonia.
13	Amides are more basic than amines.	The lone pair of electrons on the nitrogen atom are more readily available for co-ordination in amides than amines.
14	Amides can be hydrolysed by cold dilute aqueous alkalies.	$RCONH_2(aq) + OH^-(aq) \rightarrow RCOO^-(aq) + NH_3(g)$

15 The structure of a stable nitro-derivative of benzene is given by

1

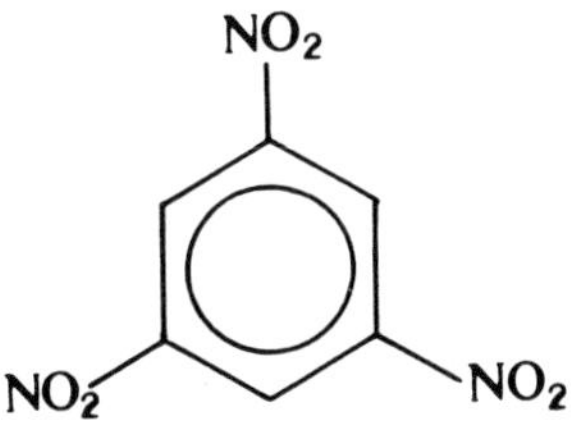

2

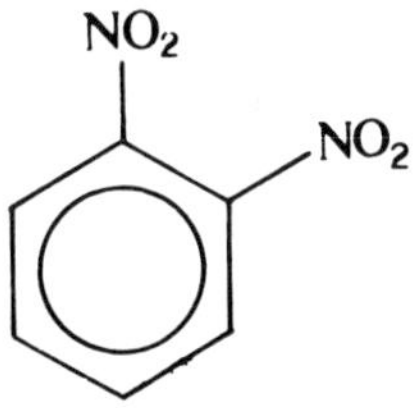

3

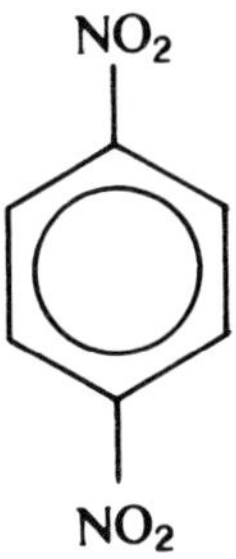

4

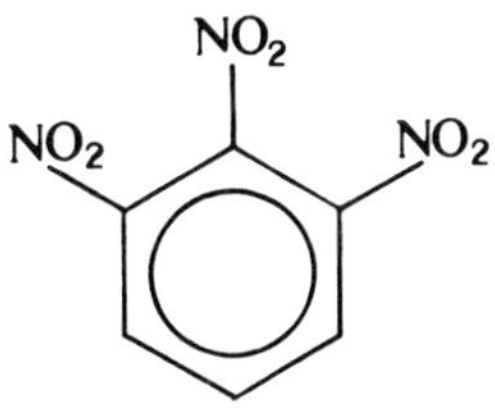

16 Which of the following reactions are undergone by aliphatic acid nitrites?

1 hydrolysis in acid conditions to yield an aliphatic acid
2 hydrolysis in alkaline conditions to yield the sodium salt of an aliphatic acid
3 reduction to an amine with sodium in ethanol
4 reduction to an amine when the vapour mixed with hydrogen is passed over a heated nickel catalyst

17 Nitroethane may be prepared by

1 the fractional distillation of iodoethane and alcoholic silver nitrite
2 the oxidation of acetonitrite with acidified potassium permanganate
3 the thermal decomposition of nitro-ethanoic acid
4 the thermal decomposition of ethanamide

18 Which of the following methods are suitable for the preparation of an acid amide?

1 heating an acid chloride with ammonia
2 heating an acid anhydride with ammonia
3 refluxing a mixture of anhydrous ethanoic acid and ammonium carbonate
4 warming an aldehyde with hydroxylamine

19 Butylamine will

1 give a vigorous exothermic reaction with an acid chloride
2 give no reaction with aqueous copper (II) sulphate
3 form hydrogen bonds with water
4 have a lower boiling point than 2, 2-dimethylethylamine

20 Which of the following types of reaction occur when an acid amide reacts with bromine and concentrated aqueous alkali?

1 elimination　　3 neutralization
2 hydrolysis　　4 rearrangement

Questions 21 to 25 refer to the list of formulae given below. Select the response which you think answers the question correctly.

A CH_2CN
$|$
CH_2CN

B $CH_3CH_2NO_2$

C $CONH_2$
$|$
$CONH_2$

D $CH_3CH_2CONHCH_2CH_3$

E CH_3CH_2ONO

Which of the above is the formula of the compound which

21 yields ammonia as a product when reduced?

22 yields hydroxylamine when hydrolysed by a mineral acid?

23 yields a compound of formula $(CN)_2$ on dehydration?

24 forms an acid of formula $(CH_2COOH)_2$ on acid hydrolysis?

25 is formed when a 1° amine is acylated with an acid amine?

18 POLYMERS, DETERGENTS, MACROMOLECULES, PROTEINS AND AMINO ACIDS

1 In which of the following reactions does glycine exhibit amine type behaviour?

A $H_2NCH_2COOH + OH^- \rightarrow H_2NCH_2COO^- + H_2O$

B $H_2NCH_2COOH + H^+ \rightarrow H_3N^+CH_2COOH$

C $H_2NCH_2COOH + CH_3OH \xrightarrow{\Delta} H_2NCH_2COOCH_3 + H_2O$

D $H_2NCH_2COOH + NaOH \xrightarrow{\Delta} H_2NCH_2COONa + H_2O$

E $2H_2NCH_2COOH \xrightarrow{\Delta}$ 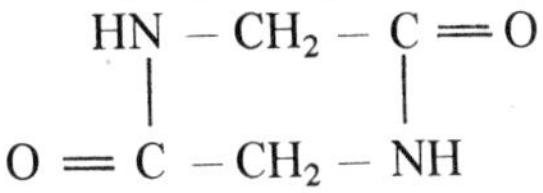 $+ 2H_2O$

2 The R_f value referred to in the separation of amino acids by paper chromatography is

A the time for the development of the chromatogram

B the ratio of the volatility of solvent and acid at room temperature

C the distance the amino acid travels in a given time

D the distance the amino acid travels relative to the distance the solvent travels

E the relative molecular mass of amino acid compared with that of the solvent

3 Which of the following is the formula of the compound which reacts with a carboxylic acid to form an ester known as a fat?

A $(CH_3CH_2CO)_2O$

B $CH_3CH_2CH_2OH$

C 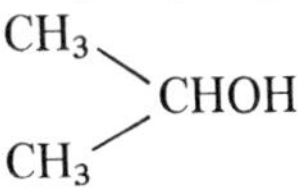

D 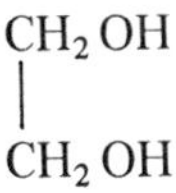

E 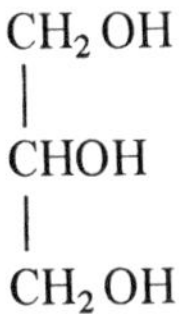

4 The constituent present in a detergent which reacts with water to provide an oxidizing agent to bleach stains is

A a soluble silicate

B sodium perborate

C a fluorescer

D a complex phosphate

E sodium sulphate

5 Which of the following is the formula of a compound which is acidic in aqueous solution?

A $H_2N\ CHCOOH$
$|$
CH_2
$|$
$CH(CH_3)_2$

B $H_2N\ CHCOOH$
$|$
$CHOH$
$|$
CH_3

C $H_2NCH\ COOH$
$|$
CH_2
$|$
CH_2
$|$
CH_2
$|$
CH_2NH_2

D
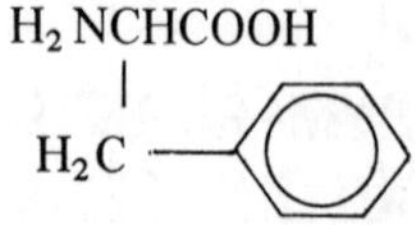

E
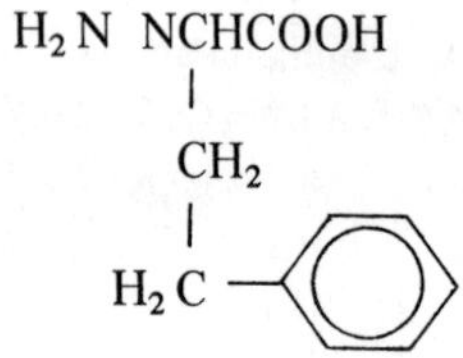

6 Nylon (66) is a polymer which can be described as a
A random copolymer
B block polymer
C minor cross-linked polymer
D linear alternating copolymer
E linear polymer

7 Which of the following methods is suitable for the preparation of a simple amino acid?
A the oxidation of an amine by acidified potassium permanganate
B hydrolysis of the product formed when a 1° amine is acylated
C refluxing a monochloro substituted aliphatic carboxylic acid with concentrated aqueous ammonia
D acid hydrolysis of the compound formed when ammonia undergoes an addition reaction with ethanal
E pyrolysing ammonium carbonate

	STATEMENT 1	STATEMENT 2
8	The functional groups present in amino acids are $-NH_2$ and $-COOH$.	When proteins are hydrolysed amino acids are formed.

9 Which of the following is the structure of glucose?

A

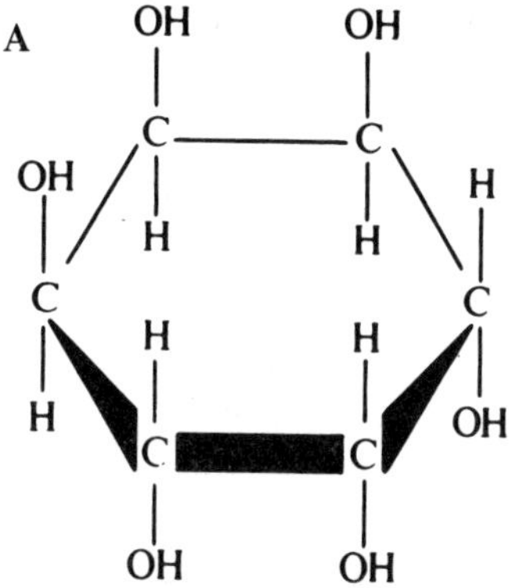

B

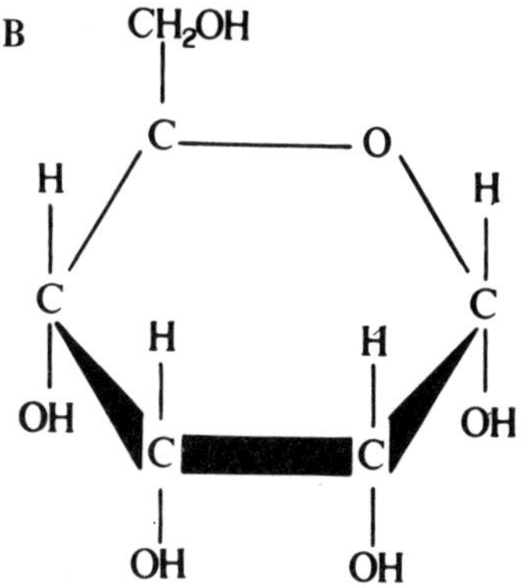

C

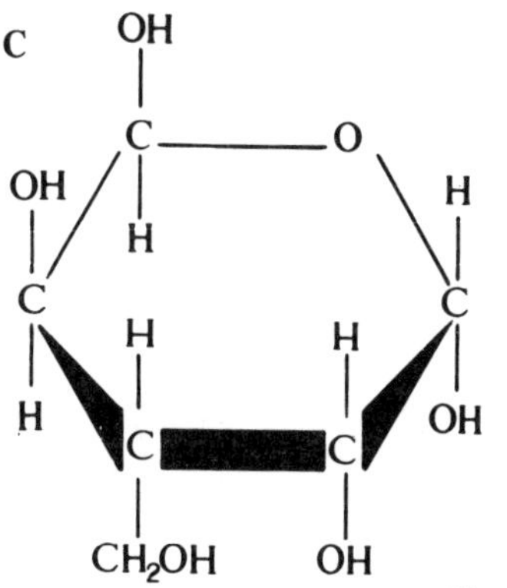

D

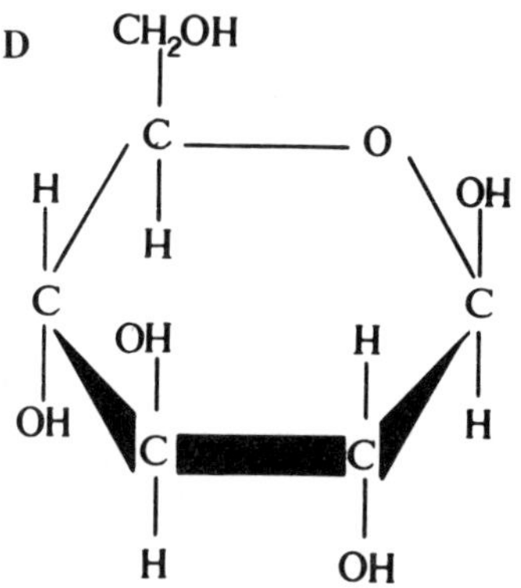

E

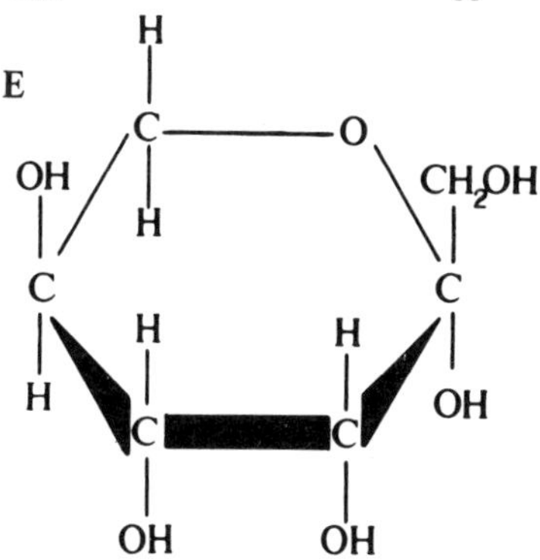

	STATEMENT 1	STATEMENT 2
10	Styrene contains a benzene ring.	The polymer formed when styrene polymerizes in the presence of benzoyl peroxide is soluble in benzene.

	STATEMENT 1	STATEMENT 2
11	Starch is a polysaccharide.	On acid hydrolysis starch yields a monosaccharide.
12	Polyvinyl chloride is a thermosetting plastic.	The structure of PVC consists of a close-linked network of long chain molecules.
13	Soapless detergents, in which the organic radical has a negative charge, are better washing agents than those carrying a positive charge.	Detergents with a negative organic radical, consisting of a straight chain alkyl group can not be degraded by bacteria.
14	It is possible for a protein molecule to have more than one conformation.	Rotation can take place within the protein molecule about the C-N bond.

15 The solution formed on neutralizing the product of acid hydrolysis of starch will

1 reduce Fehling's solution
2 give a blue colour with iodine
3 reduce ammoniacal silver nitrate to silver
4 form tri-iodomethane with an alkaline solution of iodine

16 Ethyne will undergo linear polymerization when the gas is passed through

1 hot ethanoic acid in the presence of mercury (II) sulphate
2 a very hot copper tube
3 concentrated aqueous hydrogen peroxide
4 a solution of aqueous copper (I) chloride saturated with ammonium chloride

17 The stages necessary for the conversion of iodoethane into a polymer of type $\sim - CH_2 - \overset{\overset{CN}{|}}{CH} - \sim$ are

1 reflux the alkyl halide with alcoholic alkali to form ethene
2 oxidation of ethene with hot dry silver oxide to form an oxide
3 addition of hydrogen cyanide to the oxide to form a compound

$$\begin{array}{ccc} OH & & CN \\ | & & | \\ H_2C & - & CH_2 \end{array}$$

4 elimination of water from the compound $\begin{array}{ccc} OH & & CN \\ | & & | \\ H_2C & - & CH_2 \end{array}$

18 Which of the following apply to globular proteins?

1 high solubility in water
2 contain the group –CO–NH–
3 occur in bones and skin
4 denatured on boiling with a mineral acid

19 The necessary requirements for a good detergent are

1 high solubility in water
2 ability to lower the surface tension of water
3 must be a good emulsifying agent
4 must form an insoluble salt with aqueous Ca^{2+} ions

20 Which of the following apply to mono- *and* di-saccharides?

1 may be crystallized from aqueous solution
2 their ratio of carbon to oxygen atoms is 1:1
3 have a sweet taste
4 polymerize on heating

Questions 21 to 25 refer to the formulae given below. Select the response which you think answers the question correctly.

A $(-CH_2CCl=CHCH_2-)_n$
B $(CH_2O)_6$
C $C_6H_{10}O_5$
D $(-CONHCH(CH_3)-)_n$
E $(-CH_2NHCONH-)_n$

Which of the above

21 contains a group which will give a violet colouration on the addition of one drop of aqueous copper (II) sulphate when in alkaline solution?

22 contains a group which is present in the polymer formed by the catalytic condensation of urea and methanal?

23 is present in a compound formed by simple addition polymerization?

24 undergoes condensation with a like molecule to form a macromolecule?

25 is the formula of a group present in a long chain polymer which hydrogen bonds to like long chain polymers forming fibres?

19 ISOMERISM AND RESONANCE

1 Which of the formulae given is that of a compound which exhibits stereoisomerism?

A $CH_3CH_2CH_2CH_3$
B $CH_3OCH_2CH_3$
C $CH_3CH_2CH_2OH$
D CH_3COCH_3
E $CH_3CH_2CH(CH_3)OH$

2 Ethanol and methylmethoxide are said to exhibit *functional group isomerism* because

A both compounds contain two carbon atoms
B both compounds have the same empirical formula
C both compounds have the same molecular formula and are in different homologous series
D ethanol will yield an ether on refluxing with concentrated sulphuric acid
E a methyl group can rotate freely about a π-bond in each compound

3 All of the following are resonance hydrid structures for the SO_4^{2-} ion EXCEPT

A

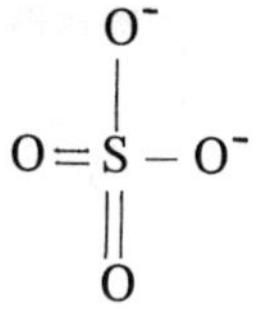

B

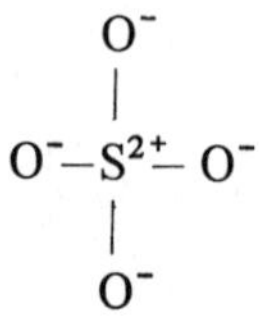

C

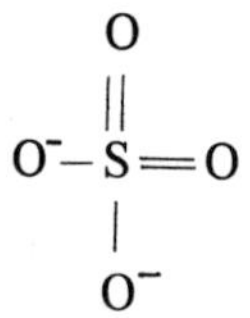

D

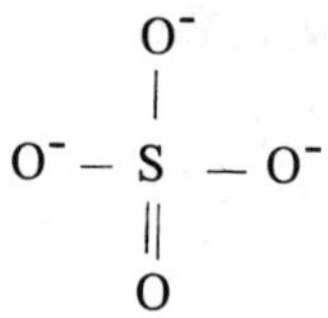

E

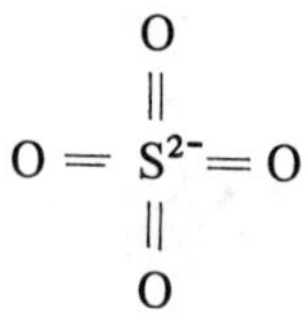

4 Which of the following can exhibit tautomerism?

A methanal
B but-1-ene
C propan-2-ol
D aminoethanoic acid
E the copper (II) ammine complex

5 The geometric isomers of the compound of formula HOOCCHCHCOOH will

A exhibit different chemical behaviour under similar conditions
B have the same degree of solubility in organic and inorganic solvents
C have two groups capable of free rotation about a carbon-carbon bond
D have different structural formulae
E melt at the same temperature

6 Which of the following is both a functional group isomer and a chain isomer of butan-1-ol?

A methylethoxide
B 2-hydroxybutane
C butan-2-ol
D butanone
E 2-methylpropan-1-ol

7 Prototropic isomerism occurs when

A a pair of isomers are in dynamic equilibrium
B a compound contains a carbonyl group
C the angle between two alkyl groups in a compound is 120°
D a keto-form of a compound is in dynamic equilibrium with an enol-form
E an aldehyde undergoes an addition-elimination reaction

8 In which of the following reactions are the principal products positional isomers?

A the catalytic decarboxylation of ethanoic acid
B the acid hydrolysis of the compound of formula

$$C_2H_5-\underset{\underset{\displaystyle NMgBr}{\|}}{C}-C_2H_5$$

C the dehydration of butan-2-ol with concentrated sulphuric acid
D the oxidation of methyl benzene with acidified potassium permanganate
E the action of heat on a mixture of 2-methylbromoethane and sodium ethoxide

	STATEMENT 1	STATEMENT 2
9	Hydrogen cyanide exhibits tautomerism.	The carbon-nitrogen bond in hydrogen cyanide contains at least one σ bond.

	STATEMENT 1	STATEMENT 2
10	A compound must contain an asymmetric carbon atom in order to be optically active.	2-hydroxypropanoic acid contains an asymmetric carbon atom.
11	There are four isomers of the ion of formula $Cu(NH_3)_4{}^{2+}$	The $Cu(NH_3)_4{}^{2+}$ ion has a planar shape with a ligand situated at each corner of a quadrilateral.
12	There are three trichlorobenzene isomers.	Chlorine induces an electromeric effect into the ring enhancing electrophilic attack.
13	Free rotation can **not** take place around the double bond between carbon and nitrogen in an oxime.	Propanone reacts with hydroxylamine to give an oxime which can exhibit isomerism.
14	The molecule of formula N_2O has a very small dipole moment even though it is asymmetrical.	The N_2O molecule undergoes resonance $N{=}N^{+}{=}O \leftrightarrow N{\equiv}\overset{+}{N}-\bar{O}$

15 Which of the following apply to 2,3-dihydroxy-butanedioic acid?

1 three isomers can be prepared
2 no isomer will have a plane of symmetry
3 an externally compensated mixture, or racemate, can be prepared
4 all isomers have the same melting and boiling points

16 A compound of formula

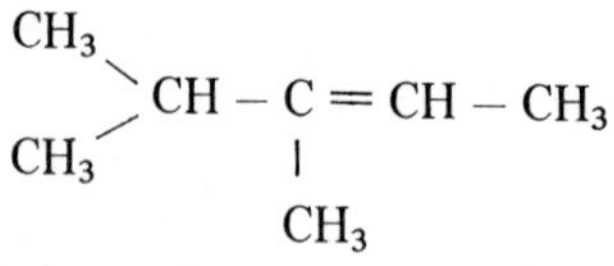

can exhibit

1 geometric isomerism
2 functional group isomerism
3 structural isomerism
4 optical isomerism

17 The degree of rotation of plane polarized light by a solution containing an optical isomer is dependent upon

1 the wavelength of the light in the optical isomer
2 the length of the path through the solution
3 the temperature of the solution
4 the concentration of the optical isomer in the solution

18 An infra-red absorption spectrum of propanone will show absorption corresponding to two isomers because

1 propanone exists in dynamic equilibrium with a tautomer
2 propanone has an asymmetric carbon atom
3 $CH_3 - CH = CH - OH \rightleftharpoons (CH_3)_2C = O$
4 the π-bond in the C=O group absorbs twice the amount of infra-red radiation absorbed by a σ-bond

19 The viscosity of the compound of formula $CH_3CH_2CH_2CH_2OH$ is greater than that of the compound of formula

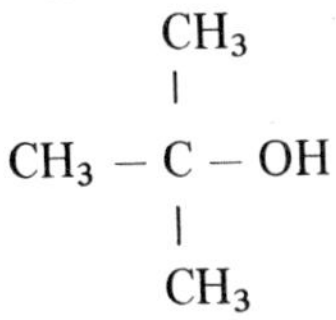

because

1 $CH_3(CH_2)_3OH$ molecules hydrogen bond with each other
2 $CH_3(CH_2)_3OH$ is a straight chain isomer
3 $CH_3C(CH_3)_2OH$ has a 3° carbon atom
4 1° alcohols are good solvents

20 In which of the following is there an inversion of configuration?

1 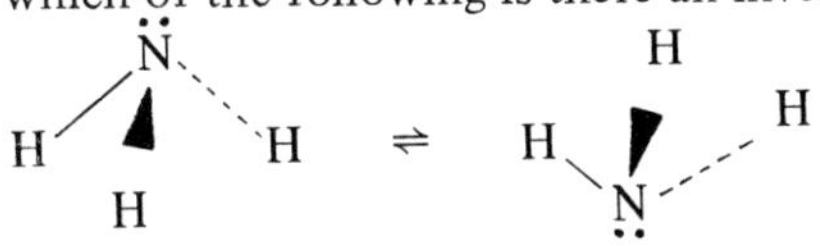

2 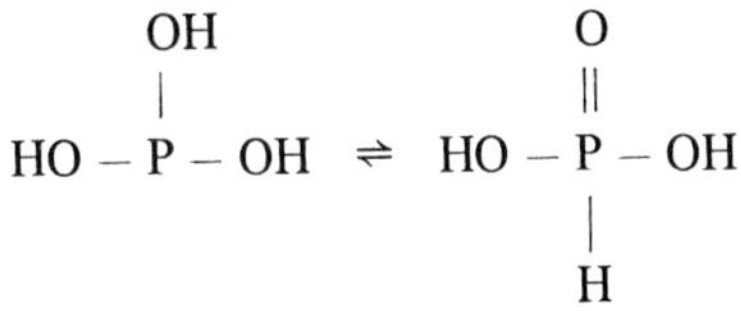

3 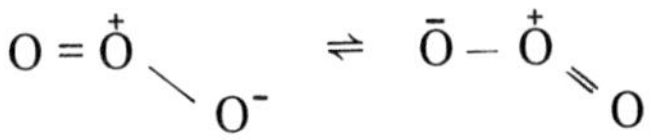

4 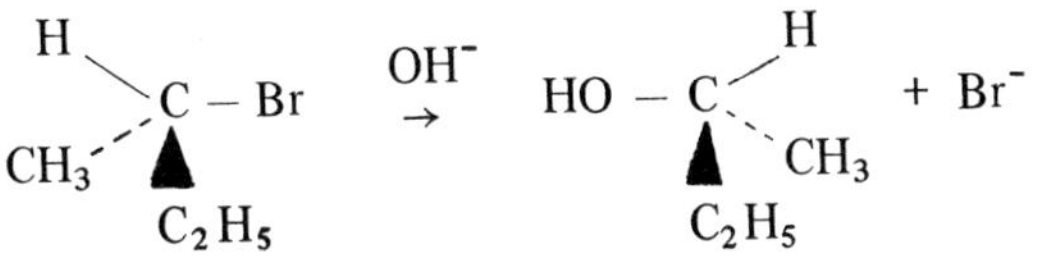

Questions 21 to 25 refer to the list given below. Select the response which you think answers the question correctly.

A optical
B geometrical
C positional
D functional group
E metamerism

Which of the above types of isomerism is associated with

21 the compounds of formulae $CH_3CONHC_2H_5$ and $C_2H_5CONHCH_3$?

22 two compounds which are enantiomorphs?

23 the ion of structural formula

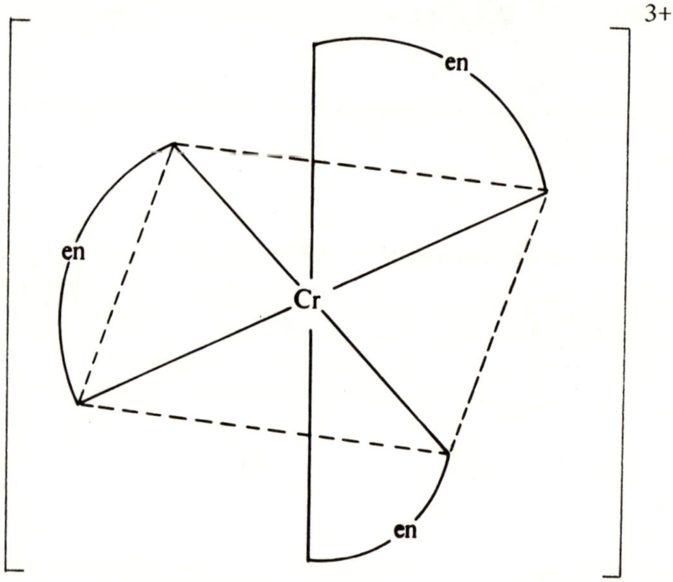

where en is 1,2-diaminoethane?

24 compounds of formula HOOC CH(OH)CH(OH)COOH?

25 the compounds of formulae CH_3CHBr_2 and $BrCH_2CH_2Br$?

ANSWERS

1

1	B	6	C	11	C	16	B	21	D
2	D	7	D	12	D	17	D	22	C
3	E	8	B	13	A	18	B	23	E
4	C	9	A	14	E	19	E	24	D
5	A	10	D	15	B	20	D	25	E

2

1	C	6	A	11	E	16	D	21	D
2	D	7	C	12	A	17	B	22	C
3	B	8	D	13	E	18	D	23	B
4	B	9	D	14	E	19	A	24	A
5	C	10	A	15	D	20	E	25	A

3

1	D	6	A	11	E	16	B	21	E
2	D	7	E	12	D	17	D	22	C
3	A	8	C	13	C	18	D	23	A
4	E	9	B	14	B	19	D	24	D
5	B	10	B	15	E	20	B	25	E

4

1	A	6	B	11	E	16	C	21	C
2	D	7	A	12	A	17	B	22	A
3	A	8	C	13	E	18	C	23	C
4	D	9	A	14	E	19	A	24	A
5	A	10	E	15	C	20	E	25	B

5

1	B	6	A	11	E	16	C	21	C
2	D	7	C	12	E	17	B	22	C
3	A	8	E	13	D	18	E	23	D
4	D	9	C	14	B	19	E	24	E
5	C	10	A	15	E	20	C	25	E

6

1	E	6	B	11	A	16	E	21	E
2	E	7	B	12	E	17	B	22	D
3	A	8	C	13	B	18	A	23	B
4	A	9	A	14	D	19	D	24	C
5	D	10	C	15	D	20	A	25	A

7

1	C	6	C	11	E	16	B	21	A
2	E	7	A	12	C	17	E	22	D
3	B	8	B	13	B	18	A	23	E
4	D	9	E	14	E	19	C	24	E
5	C	10	B	15	D	20	D	25	B

8

1	D	6	C	11	B	16	E	21	C
2	A	7	D	12	A	17	A	22	A
3	B	8	C	13	E	18	B	23	A
4	C	9	A	14	A	19	E	24	E
5	E	10	B	15	B	20	C	25	B

9

1	C	6	A	11	D	16	E	21	E
2	D	7	B	12	D	17	A	22	D
3	B	8	C	13	B	18	D	23	B
4	E	9	A	14	A	19	B	24	C
5	E	10	E	15	E	20	E	25	E

10

1	B	6	E	11	B	16	C	21	D
2	A	7	A	12	D	17	B	22	E
3	D	8	D	13	A	18	D	23	E
4	C	9	A	14	E	19	E	24	C
5	C	10	D	15	A	20	D	25	E

11

1	D	6	B	11	A	16	E	21	B
2	E	7	E	12	A	17	E	22	A
3	D	8	A	13	D	18	E	23	C
4	C	9	B	14	E	19	E	24	E
5	B	10	E	15	D	20	E	25	B

12

1	B	6	C	11	B	16	B	21	B
2	A	7	C	12	E	17	A	22	E
3	D	8	A	13	D	18	E	23	E
4	C	9	B	14	D	19	E	24	C
5	A	10	D	15	A	20	E	25	C

13

1	C	6	E	11	E	16	A	21	A
2	A	7	E	12	E	17	D	22	A
3	E	8	B	13	B	18	C	23	B
4	C	9	A	14	E	19	C	24	C
5	D	10	C	15	A	20	E	25	A

14

1	C	6	E	11	C	16	D	21	B
2	E	7	B	12	B	17	E	22	E
3	A	8	A	13	B	18	B	23	D
4	C	9	C	14	A	19	E	24	D
5	C	10	E	15	E	20	C	25	D

15

1	D	6	B	11	C	16	E	21	E
2	C	7	C	12	A	17	B	22	B
3	A	8	E	13	D	18	A	23	A
4	C	9	A	14	B	19	C	24	D
5	C	10	D	15	E	20	B	25	C

16

1	A	6	C	11	B	16	E	21	A
2	A	7	A	12	C	17	A	22	E
3	E	8	B	13	D	18	B	23	A
4	D	9	A	14	A	19	D	24	A
5	C	10	E	15	E	20	B	25	B

17

1	C	6	E	11	C	16	E	21	E
2	A	7	A	12	A	17	B	22	B
3	A	8	E	13	E	18	A	23	C
4	B	9	D	14	D	19	B	24	A
5	A	10	A	15	A	20	A	25	D

18

1	B	6	E	11	B	16	D	21	D
2	D	7	C	12	E	17	E	22	E
3	E	8	B	13	C	18	C	23	A
4	B	9	E	14	A	19	A	24	B
5	E	10	B	15	B	20	B	25	C

19

1	E	6	A	11	D	16	B	21	E
2	C	7	D	12	B	17	E	22	A
3	E	8	C	13	D	18	B	23	A
4	D	9	B	14	A	19	E	24	A
5	A	10	E	15	B	20	E	25	C